IN THE SKIN OF A JIHADIST

IN THE SKIN OF A JIHADIST

A YOUNG JOURNALIST ENTERS THE ISIS RECRUITMENT NETWORK

ANNA ERELLE

TRANSLATED FROM THE FRENCH BY ERIN POTTER

HarperCollins*Publishers*

HarperCollins books may be purchased for educational, business, or sales promotional use. For information, please e-mail the Special Markets Department at SPsales@harpercollins.com.

Originally published in France in 2015 by Éditions Robert Laffont under the title *Dans la peau d'une djihadiste.*

FIRST EDITION

Library of Congress Cataloging-in-Publication Data

Erelle, Anna.
[Dans la peau d'une djihadiste. English]
In the skin of a jihadist : a young journalist enters the ISIS recruitment network / Anna Erelle ; translated from the French by Erin Potter. — First edition.
pages cm
"Originally published in France in 2015 by Éditions Robert Laffont under the title Dans la peau d'une djihadiste."
ISBN 978-0-06-241707-7 (pbk.) — ISBN 978-0-06-241709-1 (ebook)
1. Erelle, Anna. 2. IS (Organization) 3. Terrorists—Recruiting—France. 4. Women terrorists—France. 5. Terrorism—France. 6. Journalists—France. I. Title.
HV6433.F7E7413 2015
956.05'4092—dc23

2015010342

15 16 17 18 19 OV/RRD 10 9 8 7 6 5 4 3 2

For Éric and Noël
For Pauline and Jérôme

Only voluntary, inspired self-restraint can raise man above the world stream of materialism. . . . Even if we are spared destruction by war, our lives will have to change if we want to save life from self-destruction.

Aleksandr Solzhenitsyn, "A World Split Apart,"
Harvard University commencement address, June 8, 1978

The following events took place during the spring of 2014, two months before the Islamic State occupied Mosul, the second-largest city in Iraq, and its leader, Abu Bakr al-Baghdadi, declared a worldwide caliphate.

"Listen to me! I love you more than I've loved anyone. You should be here with me. I can't stand to think of you in that corrupt country. I'll protect you. I'll shelter you from the world's evils. When you come to live with me, you'll see what a paradise me and my men are building. You'll be amazed. Here, people care about each other. They respect each other. We're one big family, and we've already made a place for you—everyone is waiting for you! You should see how happy the women are here. They used to be like you—lost. One of my friends' wives has arranged a program for your arrival. After your shooting lessons, she'll take you to a very beautiful store, the only one in the country that sells fine cloth. I'll pay for everything. You'll establish your own little world here with your new friends. I'm so excited for you to be here. Mélodie, my wife! Hurry up; I can't wait."

Mélodie stares into her computer screen, admiring the strong man eighteen years her senior. She loves him, even if she's only ever seen him on Skype.

"Do you really love me?" Mélodie murmurs, her voice childish and frail.

"I love you for the sake of Allah. You are my treasure, and the Islamic State is your home. Brick by brick, we'll build a better world, a place where *kafirs*[*] won't be allowed, and we'll carve a name for ourselves in history. I've found a huge apartment for you! If you bring friends, I'll find an even bigger one. You'll take care of orphans and the wounded during the day, while I'm fighting. We'll spend our evenings together . . . *insha'Allah*[†]."

Mélodie feels loved. She feels useful. She's been looking for purpose in her life: now she's found it.

* *Infidels, in Arabic. Footnotes are the author's except those indicated as the translator's.*

† *Translator's note:* insha'Allah *is Arabic for "God willing." It is used rather ubiquitously in the Arabic-speaking world in the context of future hopes and plans.*

Paris, ten days earlier

I was frustrated that Friday night as I left the editorial offices of a magazine where I do freelance work. The paper had received a letter from a lawyer forbidding me from publishing an article I'd written about a young female jihadist. I had just spent two days in Belgium with Samira, the girl's mother. Her daughter ran away to Syria a year before to join Tarik, the man of her life and a fanatic devoted to the Islamic State's cause. Naïve and blind with emotion, Leila* wanted to live with her great love. A bullet to the heart ended his twenty years and one spring. Samira was hopeful when she learned of the death of the man she'd been forced to consider her son-in-law. With Tarik dead, Samira saw no reason for her daughter to stay in the tragically war-torn country, but Leila was clear: she now belonged to that sacred land and wanted to do her part in the fight to create a religious state in the Middle East. With or without her husband. Tarik had been an

---

** Her name has been changed.*

emir,* which meant his widow was well taken care of. People respected her, and Leila asked her mother, "Why should I go back?"

Local news sources had picked up the story and begun comparing the eighteen-year-old jihadist to the black widow, a prominent figure in the world of international terrorism and the wife of the man who assassinated Ahmad Shah Massoud.† Samira's love for her daughter was great, and her response to the situation swift, but she was coming up against an immense challenge. Not only did she have to find a way to repatriate Leila to Belgium; she also had to prove to the authorities that her daughter was living in one of the most dangerous countries on earth for humanitarian reasons. Otherwise, Leila would be considered a threat to domestic security and sent to prison, before potentially being banned from setting foot in her own country.

That was when Samira's and my paths crossed. Journalism can lead a person to many things, sometimes to the aid of a distressed mother. Samira was beside herself, and she'd turned to Dimitri Bontinck, a former member of the Belgian Special Forces who famously managed to repatriate his own son from Syria. Dimitri is a source of hope for all these European families who wake up one morning to the

---

** A title of nobility used throughout the Muslim world. The Islamic State is an extremely hierarchical organization that ranks its followers. Generally, an emir earns his stripes through determination, strength, and faith, which he proselytizes in the face of opposition. Emir is one of the organization's most prestigious honorific titles.*

† A hero in the struggle against Soviet occupation in Afghanistan, Massoud could have led the coalition against the Taliban had he not been assassinated on September 9, 2001, two days before the attack on the World Trade Center and the Pentagon.

harsh realization that even those they'd least suspect, even their own teenagers, could be jihadists. After his personal experience, Dimitri became a tireless crusader, volunteering for virtual suicide missions to save other youths—or at least dig up concrete information to help their families. Aware of the risks that Leila faced for being branded the "new black widow," he'd asked me to meet her mother. I'm a journalist, and though I'm keenly interested in geopolitics, I'm not an expert. However, I've always been drawn to erratic behavior, whatever the cause—religion, nationality, social milieu. I'm fascinated by what motivates people to make fatal decisions. Sometimes it's drugs. Sometimes it's crime or marginality. I've also done a lot of work on radical Islam. Back then, I'd been studying European jihadists in the Islamic State for about a year. There were many similarities between the successive cases, but I was interested in understanding what it was that made each individual decide to give up everything and brave death for this cause.

At the time, Dimitri and I were writing a book about the nine horrifying months he spent looking for his son. We spoke with many European families facing the same ordeal. I tried to interview as many people as I could. I saw the impact of digital propaganda on God's newly minted soldiers, but I still didn't understand what drove them. Why did they leave everything—their past, their families? Over the course of a few weeks, they threw away their lives, convinced they'd never look back. Ever. Walking through their bedrooms, often preserved by their parents, always gave me chills. I was peering into other people's intimate spaces, which had become shrines to forgotten lives, as if their teenage relics were the last proof of their existences. Leila's existence seemed frozen in time. Pictures of her "normal" life abounded. There she was in a tank top, wearing makeup, at friends' houses, or in a café. These idealized images were a far cry from the new Leila with her burqa and her Kalashnikov.

After listening to Samira's story, I continued my investigation, which confirmed some of what she'd told me, and I wrote the article. Yet another piece on a subject that had become increasingly ubiquitous over the past several months. But it wouldn't be published. Leila was furious when her mother mentioned our interview, and threatened to burn all bridges. "If you talk about me to the press," her panicked mother tearfully reported her words, "not only will I never come back, you'll never hear from me again. You won't know if I'm dead or alive." After that, I couldn't convince the mother to let me publish. In absolute terms, I didn't need her permission to do it—the story was already public knowledge in Belgium. But what good would it do? Sadly, each week brimmed with new stories like this one. I was all-too familiar with the determination of these young people who believed they'd found faith. All day, they were bombarded with messages to forget their "depraved" families and open their arms to their new brothers. "Infidels," even if called "mom" or "dad," were seen as obstacles in their spiritual journey.

It wasn't Leila's fault. She honestly believed she was protecting her mother by telling her how to behave. Alone at home, I got worked up over the methods of propaganda used by Islamists. Searching for videos of Tarik alive, I came across an incalculable number of propaganda films on YouTube. I muted the sound whenever the language wasn't French or English. The monotonous chants went to my head, deadening my mind. I couldn't listen to them anymore. Still, the sounds were more tolerable than the images of torture and charred bodies laid out in the sun. Wandering through jihadist Francophone networks online, I was continually shocked by the contrast between sound and image. The juvenile laughter accompanying these horrific scenes made the videos all the more unbearable. I'd noticed an uptick in activity over the past year. Many teenage jihadists have a second Facebook account, registered under a fake identity. They act normal around their families, but once alone in their bedrooms, they travel to their virtual world, which they take for reality. Some call for murder, though without really understanding the impact or significance of their messages. Others encourage jihad. Girls share links about Gazan children, underscoring the suffering of the very young. The girls' pseudonyms all begin with *Umm*, "mom" in Arabic.

Social networks contain precious information for those who know how to look. That is why, like many other journalists, I had a fictional account I'd created several years before. I used it to keep an eye on current events. I rarely posted on the account, and when I did it was very brief, and only directed at my list of approximately one hundred "friends" from around the world. My name on this account was Mélodie. My followers weren't using their real identities, either. Avatars ensure anonymity, which allows users to express themselves more freely and accounts for the growing number of young people attracted to Islamist propaganda.

New technologies have of course bred new forms of proselytism. I spent hours scanning users' public descriptions of gruesome or simply outrageous plans. Happily, not all of the teenagers writing about criminal activity become murderers. For some, Jihadism 2.0 is a fad. For others, it represents the first step on their path to radicalism.

I spent that Friday night in April on my couch, stewing over the gag order on my article and flicking from account to account. Suddenly I came across a video of a French jihadist who looked to be about thirty-five. The video showed him taking inventory of the items inside his SUV. It was like a bad parody of the farcical news show *Les Guignols de l'info.* I smiled wryly at the deplorable images. I wasn't proud of myself, but I couldn't help watching; it was absurd. The man in the video wore military fatigues and called himself Abu Bilel. He claimed to be in Syria. The scene around him, a true no-man's-land, didn't contradict him. He proudly brandished his CB radio, which looked like it came straight out of the 1970s. He used it to communicate with other militants when he couldn't reach them through telephone networks. In reality, it crackled more than it communicated. In the back of his car, his bulletproof vest sat beside one of his machine guns, an Uzi—a historic gun originally manufactured for the Israeli military. He presented a series of weapons, including "an M16 stolen from a marine in Iraq"—I burst out laughing. The factoid, I would later learn, was entirely plausible. I would also discover that Abu Bilel was not as stupid as he seemed. In fact, he had spent the past fifteen years waging jihad all over the world. But for the moment I knew nothing of the bellicose man on my screen proudly unveiling the contents of his glove box—a thick stack of Syrian pounds, candy, a knife. He removed his reflective Ray-Bans, revealing darkly lined black eyes.

I knew that Afghani soldiers used eyeliner to keep their eyes from tearing up when exposed to smoke. Still, seeing a terrorist with eyes made up like my own was surprising, to say the least. Abu Bilel spoke perfect French, with what sounded to me like a very slight Algerian accent. He smiled broadly in an expression of self-satisfaction as he beckoned viewers and called for *hijrah*.*

I shared his video. I usually kept a low profile on my account, but I occasionally imitated my digital peers in order to carve a place for myself in their world. I didn't preach or encourage the cause. I simply posted links to articles relating strikes by Bashar al-Assad's army or videos like this one. My profile picture was a cartoon image of Princess Jasmine from the Disney movie *Aladdin*. For my cover photo, I uploaded a popular slogan I'd seen online: "We'll do to you as you do unto us." I tended to change my profile location depending on whatever story I was presently researching. Now I claimed to be in Toulouse, a city in southwestern France. Over the past five years, many stories had led me there, notably, the shooting carried out by Mohammed Merah in 2012. The housing project where he'd lived in the northeastern outskirts of Toulouse was an endless mine of information. It was also an important hub for the traffic of hashish.

---

* Hijrah *refers to the journey of the Prophet Muhammad and his followers from Mecca to Medina in the seventh century. Today, it is used more generally to mean leaving a land of unbelievers (*kafir*, in Arabic) to join an Islamist country.*

I was actually in Paris, casting around for a fresh angle on the departures to Syria. Many of these tragic cases resembled one another, and I suspected that readers were saturated with information. In addition, the nightmarish situation in Syria made it difficult to analyze. Each week, I worked with editors, trying to find new angles. Each week, we arrived at the same conclusion: would-be jihadists came from all sorts of social backgrounds and religions; they turned to radical Islam after a single failure or a lifetime of not fitting in; then they left for Syria to join one of the many Islamist gangs that have been proliferating there. Yes, but despite the similarities, after having spent so much time working on these issues, I had grown attached to individual families. I cared about their children and their stories, even if I wasn't likely to meet them. And I had actually met some "teens" drawn to jihadism while I was working on stories. Today, when I see them again, they tell me they want to go there. There? "What's there for you?" I ask them, exasperated, "Except death and the opportunity to become cannon fodder?" The response is almost always the same: "You don't understand, Anna. You're thinking with your head, not with your heart." I exhaust myself coming up with dubious comparisons to historic events. Germany, a country rich in culture, fell into Hitler's hands during the last century. Or the black-and-white view of the world according to communism. Or the generation of 1970s intellectuals who extolled the virtues of Maoist thought, insisting that truth resided in the *Little Red Book*. But my cyber interlocutors poke fun at my historical references, pointing out that red and green are very different colors. However, I'm not talking about the Koran, which has nothing to do with fanatic ideology.

In 2014, journalism was no longer a respected profession. And when one worked on "societal" issues, it was out of passion. If only I could write about this topic in a new way, one that avoided treating individuals as part of a succession of similar cases. I wanted to investigate the roots of "digital jihadism" and get to the bottom of an evil phenomenon affecting more and more families—of all religious backgrounds. To dissect how kids here fell into the trap of propaganda, and to grasp the paradox of soldiers there who spent their days torturing, stealing, raping, killing, and being killed, and their nights staring into their computers and bragging about their "exploits" with the maturity of video-game-obsessed preteens.

Deep in reflection, I was feeling discouraged but unwilling to give up, when my computer alerted me to three messages sent to "Mélodie's" private inbox from . . . Abu Bilel. It was surreal. There I was, at ten o'clock on a Friday night in spring, sitting on my sofa in my one-bedroom Parisian apartment, wondering how to continue my investigation on European teenagers tempted by Islamic extremism, when a French terrorist based in Syria all of a sudden started writing me. I was speechless. At that moment, the only thing of which I was certain was that I hadn't imagined starting my weekend like this.

The same night

"*Salaam alaikum*, sister. I see you watched my video. It's gone viral—crazy! Are you Muslim?"

"What do you think about mujahideen?"

"Last question: are you thinking about coming to Syria?"

He certainly got straight to the point! I didn't know what to do. I instantly understood that speaking with this jihadist offered a unique opportunity that might lead to a mine of information, and I was eager to respond. When you present yourself as a journalist, it's difficult to get people to speak sincerely. In this case, my interlocutor didn't know who I was. Using this account to request information for an article didn't bother me. However, the idea of starting a conversation with a person who didn't know who I was introduced an ethical problem. I took five minutes to think. Long enough to consider his code of ethics . . . and then I replied:

"*Walaikum salaam*. I didn't think a jihadist would talk to me. Don't you have better things to do? LOL. I'm not prejudiced against fighters. Anyway, it depends on the person."

I also told him I'd converted to Islam, but didn't offer any details. I deliberately included spelling mistakes, and I tried to use teen vocabulary—LOL, LMAO, ROFL,* and other acronyms they pepper into their correspondence. I waited for his reply, a knot in my stomach. I wasn't afraid; I just couldn't believe this was happening. It seemed too big to be true. I'd interviewed mujahideen before, but never anyone over twenty years old, and never anyone who expressed anything outside of the official propaganda. While I waited, I surfed the Web, scanning other pages. Barely three minutes had passed when my computer alerted me to a new message.

"Of course I have a lot of things to do! But here it's eleven o'clock at night and the fighters are finished for the day. Do you have any questions about the video you shared? I can tell you about everything going on in Syria—the only real truth: Allah's truth. We should talk over Skype. I'll give you my username."

Bilel was direct . . . and authoritative. Skype was out of the question! I ignored his proposal and suggested we talk another time. Mélodie wasn't available now. Abu Bilel understood; he didn't want to bother her. He'd make himself available for her tomorrow whenever she wanted.

"Tomorrow?" I asked, surprised. "Are you sure you'll have Internet access?"

"Of course. I'll be here. I promise." Then, a minute later:

"You converted, so . . . you should get ready for your *hijrah*. I'll take care of you, Mélodie."

First Skype, now *hijrah*! Abu Bilel didn't lose any time!

** LOL: laughing out loud; LMAO: laughing my ass off; ROFL: rolling on the floor laughing. [Translator's note: these acronyms are English equivalents of the French acronyms used by the author in her correspondence with Bilel.]*

This was our first encounter. We'd only exchanged a few lines. He didn't know anything about this girl, except that she'd converted to Islam, and he was already asking her to join him in the bloodiest country on earth. He was shamelessly inviting her to abandon her past, her home, and her family—that is, unless they perhaps wanted to join her on her spiritual journey? He was asking her to be reborn in a new land and wait for God to open his doors to her. After an initial shock, I felt a mix of feelings. I had trouble distinguishing them all, but I was sure of one thing: I was disgusted. Bilel was targeting the weak, and whenever they took his bait, he and others like him from the Islamic State tried as hard as they could to reformat them, erasing their pasts as one would clean up a disk before recording new information. Thinking about this process and the girls he preyed on infuriated me. Going after a girl like Mélodie was so easy—and so unfair. I'd met a thousand girls like her. They hadn't had stable upbringings. Nor had they received a proper education. They didn't have any guidance, so they were prone to believing rumors. It was the same for boys. It made me so angry, I wanted to punch him.

What was I getting myself involved in? I sensed it would go much further. But I never imagined that six months later, at the present moment of writing, Abu Bilel would continue to impact my life. For the time being, all I could think of was the fact that if I wanted to glean information from this terrorist, Mélodie would really have to exist. As in spy stories, I needed to craft a story for her. She would step through the looking glass, and perhaps even be sacrificed in the end. I would give her traits from all the kids I'd met who'd succumbed to jihadism. She would be a melting pot of Norah, Clara, Leila, Élodie, the Bon brothers, Karim, and Karim's best friend. Their families had to go to the border between Turkey and Syria just to obtain proof that they were alive. Most returned empty-handed. If Mélodie began corresponding with this man, who seemed experienced, given his age, perhaps he would reveal useful pieces of information. Nothing ventured, nothing gained. Besides, I had too many unanswered questions. Any answers I could obtain would be precious for future stories. I undertook this project as an anthropological study. For now, however, it was getting late, and I wanted to stop thinking about Abu Bilel.

My boyfriend was due to arrive. I called to tell him I wanted to spend the night at his apartment. I didn't say anything about how I'd spent the evening, only that I wanted to sleep next to him.

Saturday morning

Milan handed me a Diet Coke, *M*, which is a weekly magazine published by *Le Monde*, and his iPad. Coke is my morning coffee; I still haven't learned to drink grown-up beverages at their designated times. Milan is familiar with my routine, and his tablet is always connected to Mélodie's Facebook account. That way I can keep an eye on her News Feed. While we were sleeping, Abu Souleyman,* a young Alsatian in Syria, died. A picture of his body, a faint smile on his lips, was being widely shared and commented on by dozens of users. Milan cuddled up to me and sipped his coffee. He looked at me tenderly, shaking his head. "Is this going to take long?" he asked, still half asleep. I smiled and kissed him. He flipped through a magazine on French cinema while I scanned the day's comments on the "martyr." Nothing original. Apparently Souleyman was in a better place. God was proud of him. And we should be "proud he died for his cause" at the age of twenty-one.

---

* *His name has been changed.*

Other conversations interested me more. Abu Bakr al-Baghdadi, the leader of the Islamic State, was said to have almost fallen into an ambush by Jabhat al-Nusra. The al-Nusra Front is one of the principal armed terrorist groups associated with al-Qaeda in Syria. This group is often wrongly conflated with ISIS, the Islamic State of Iraq and al-Sham. Although the groups' relations were once cordial, and even harmonious, that is no longer the case. Their goals and adversaries are not the same. Al-Qaeda's enemy continues to be the West—those of the cross. ISIS seeks to create an Islamic State, a Sunni caliphate somewhere between Iraq and Syria. ISIS's aim is to eliminate from power all those directly or indirectly related to Shiites, starting with the minority Alawite branch, which runs the country, before dislodging Shiite power in Iraq. A return to the Middle Ages, a triumphant Islam, and territories seized through force: those are the methods and aims of the Islamic State. Al-Qaeda shares this same ideology but first seeks to diminish Western power and demonstrate its own force, as in the attacks of September 11, 2001. To simplify matters, ISIS seeks to eliminate heretics from its geographical area, while al-Qaeda targets infidels.

Whenever my interviews lead me to a jihadist, I question him on his ambitions in the event the organization manages to achieve its goals and conquer the Middle East. I usually get the same story: "The Islamic State will wage war on the United States and force its people to submit to God's will. Then we'll abolish all borders, and the earth will be one Islamic State under sharia law." By creating a territorial seat for its utopia, ISIS has succeeded where al-Qaeda has failed. While the latter has tediously built cells throughout the world, ISIS has waged war, implemented real policies, and grown an army of fanatics—officially in Syria and unofficially in Iraq. ISIS's army first consisted of Sunnis hostile to the American invasion of Iraq; later thousands of foreign fighters swelled the ranks. Meanwhile, the terrorist organization has also refined its favorite weapon: digital propaganda. The image of Afghanis in caves doesn't entice. Jihadism 2.0's new communication strategy has hit the mark. The Islamic State has inundated YouTube with ultraviolent videos that stick in the minds of thousands of Westerners lobotomized by the group's swiftness of action and execution of threats. Promises bind only those who listen to them, it is said. Sadly, that truism especially applies to these young jihadists. Desperate for attention, the majority leave for the front with the ultimate ambition of posting pictures online of themselves dressed as soldiers. There they are sure to be noticed and feel important, and they can also express their exploits over Facebook and Twitter. Andy Warhol's 1968 pronouncement, "In the future, everyone will be world-famous for fifteen minutes," has never been so apt.

I was born in the early eighties. Religion was already an issue for teens back then, but it didn't motivate them to act the way it does today, even if some young men did become jihadists. These days, would-be jihadists aren't drawn by easy money, guns, or drug dealing. Instead they dream about being respected and gaining recognition. They want to be "heroes." Becoming the neighborhood big shot and hanging out over PlayStation is one thing; playing war and creating a state is quite another. Still, there's more than one type of jihadist. Recent cases of young people moving to the Middle East have often involved solitary radicalization. I'm thinking of a young girl from Normandy who thought she'd found the answers to her existence, alone, on the Net. A few weeks later, the converted Christian girl left to join the ranks of Islamist fighters. I imagined my Toulousain avatar, Mélodie, to be an extremely sensitive girl; being dominated would give her life a sense of purpose. Like so many other young people—from throughout history and regardless of social milieu—she lived a life of desperation.

That night

Milan was asleep. His bedroom was calm and quiet. I tossed and turned. The blinds were open, the streetlamps outside bathing the room in a poetic light. This familiar nocturnal scene accompanied my insomnia but did nothing to silence the questions crashing through my brain.

I carefully got out of bed. Milan slept like an angel, while my subconscious dragged me into the living room and toward a demon imprisoned behind a retina display. Three new messages from my correspondent awaited me. I hadn't expected so many. I lit a cigarette. He'd sent the first one at 2:30 p.m. in Syria, a surprising time for a zealous fighter to be corresponding. He should have been on the front. Or elsewhere. I was bewildered by the thought of him digitally stalking a girl from an Internet café in the middle of the afternoon.

"*Salaam alaikum*, sister. How are you today? I wanted to let you know that I'm available if you want to talk. I'm around."

Around? Around where? His next message grabbed my attention before I could reflect on that question:

"What time will you be online? I really want to talk to you."

"I have a special surprise for you . . . *Masha'Allah*[*]."

The "surprise" was a picture of him, armed to the teeth. So cool. A gigantic M4 assault rifle was slung across his shoulder. A black bandana embroidered with the Islamic State's white insignia covered his forehead. He stood erect, puffing out his chest, smiling. I had trouble believing this was real. He didn't know me. What if I was hiding behind Mélodie's identity? What if I was really a cop? Or a journalist searching for reliable information from a solid source? Abu Bilel wasn't concerned. Clearly, he thought he'd caught a fish. Based on the tone of his messages, it didn't seem like he was going to let this one escape from his net. Did he often act like this? It must have been four o'clock in the morning. I was looking for answers. For now, all I had were more and more questions.

People often compare journalists to dogs in search of bones to gnaw on. Admittedly, at that moment, I was excited by the idea of delving into the mind of an assassin—this assassin. I admire people of faith. I envy the strength it affords them. Faith is a precious source of support as one confronts life's inevitable difficulties. But when people use spirituality as an excuse to commit murder, I, Anna, give myself permission to become someone else. At least digitally speaking. That was how I justified becoming Mélodie, a desperate and naïve young woman. Some might object to my methods on

---

* *Translator's note:* Masha'Allah *is an Arabic phrase meaning "God has willed it," and is used to express gratitude for an event or person.*

moral grounds, but at the time this terrorist organization was doing everything in its power to enroll a maximum number of new recruits. I let my conscience decide. Abu Bilel wouldn't be the subject of a story. I wanted to examine what he said and untangle fact from fiction. How many people now served the Islamic State? How many French? How many Europeans? Did women really pleasure jihadists as a way of serving God? Did they also take up arms? Abu Bilel beckoned me onto his path of religious domination, while he decimated the meek and helpless in a country rife with religious divisions. Could I get him to tell me about the bloody conflicts he spearheaded?

As day broke, I surfed the Net, scanning the labyrinthine Web for anything I could find on Abu Bilel. I dug up dozens of conversations between mujahideen and potential recruits. Nothing conclusive. However, I learned that a very important battle had just taken place in Syria, in the region of Deir ez-Zor, less than three hundred miles from the border with Iraq, a country still haunted by the ghost of Saddam Hussein and the American invasion. I came across an exchange that normally would have interested me: "We destroyed them! I recorded the whole thing! But al-Baghdadi and his emirs were suspicious it might be an al-Nusra trap, and they stayed inside the house. Call Guitone; he's with them." I'd known of al-Baghdadi, the very dangerous leader of ISIS, for a long time. But that night, since I couldn't find any information on Bilel, I was interested in Guitone. I knew him "well." Guitone, aged twenty-two or twenty-three, was born in Marseille and had lived for a while in Great Britain before joining ISIS, where he quickly climbed the ranks. He possessed three qualities that made him an essential asset to the Islamic State's digital propaganda campaign: he was good-looking, he knew his religion by heart, and he was able to preach in four different languages.

My colleagues and I had nicknamed him "the Publicist." Whenever we needed information, we could rely on him. He was always eager to help. Guitone knew me through my true identity: Anna. We had spoken on several occasions. I'd last contacted him in March about Norah, a fifteen-year-old girl from Avignon. Her family had recently told me that Norah had left to join the al-Nusra Front, and not the Islamic State. Guitone had confirmed that fact as well as her geographic location.

Guitone bragged about his affiliation with ISIS on his Facebook page, often posting videos of himself: Guitone visiting wounded jihadists in hospitals; Guitone flouting France and Turkey, armed to the teeth at a feast on the Turkish border; Guitone waving to a crowd of fighters celebrating in the conquered streets of Raqqa, Syria. Guitone was unbelievably famous. Each of his posts literally made adolescents from all over Europe salivate. He claimed to live like a king, and he was always dressed from head to toe in name brands. He was respected for what he was. He always had an innocent smile on his face. That was his trademark. Who better to convince you to embrace his cause, particularly in a country so affected by war? Admittedly, it was clever PR. I considered sending Guitone a message asking him to fill me in on the latest battle, at which the "emirs" were nowhere to be found, but I decided against it. I didn't yet know that Guitone, Abu Bilel, and al-Baghdadi were related—insanely related. I continued dissecting the information at my disposal. I had nothing on Bilel. Who was he? And how old? I guessed he had extensive experience in the field. My curiosity growing, I sensed this man was more complicated than the young jihadists I'd encountered before.

Sunday night

"Sympathy for the Devil" by the Rolling Stones crashed against the walls of my living room, resonating like a premonition. I turned on my computer and found new messages from Bilel. I barely had time to read them before he connected and contacted my digital puppet. In his first posts, he struggled to hide his crass insistence. Every other line, the mercenary begged Mélodie to sign off Facebook and continue her conversation with him over Skype, a platform that combines sight and sound. Why was he so obsessed? Was it a safety measure? Did he want to verify my identity? Or did he want to make sure the new fish swimming in his net was appetizing?

"Why do you want to Skype?" I had Mélodie reply awkwardly.

"Conversations over Skype are more secure, if you see what I mean."

No, I didn't see. He ended his sentence with a smiley face, a yellow, winking emoticon. It was absurd. He was absurd. On his profile, he swore he was "devoted to the Islamic State," so I tried to engage him on that point.

"I see you work for the Islamic State. What's your job? In France, people say it's not a very strong brigade."

I couldn't help using Mélodie to insult him. I also added a blushing smiley face. Bilel was quick to defend his vanity, firmly insisting that ISIS embodied the height of power, not only in Syria but throughout the world. Soldiers came from all corners of the globe to join its ranks.

"There are three types of fighters," my charming interlocutor went on, in teacher mode: "those on the front, those who become suicide bombers, and those who return to France to punish infidels."

"Punish? How?"

"You know how . . . like Mohammed . . ."

It was a reference to Mohammed Merah, the shooter in Toulouse. But Mélodie didn't understand.

"Who's Mohammed? And how is he punishing people?"

"You live in Toulouse, right? You don't know about the scooter killer? . . . There's one important rule: terrorize the enemies of Allah."

"But Merah killed children. Don't children represent innocence and purity? How can they be enemies?"

"You're so naïve, Mélodie. . . . You like children? One day, you'll have some of your own, *Insha'Allah.* You know, we have many orphans here in need of mothers. ISIS sisters take care of them; they're remarkable women. You have a lot in common with them. You would like them."

Although he didn't know Mélodie, Bilel was a master manipulator. His method: lull her into a state of security by telling her what she wanted to hear. Ultimately, the subject of conversation didn't matter; he would guide her in whatever direction he wanted. Mélodie expressed a certain affection for children, so Bilel suggested she could become a surrogate mother. Forgetting the discussion about Mohammed Merah, she smiled faintly, and imagined what it would be like to devote herself to others worse off than herself. As if other people's despair could cure her of her own. For some time, she'd felt lost in her depressing surroundings. Everything seemed like a waste of time. Nothing mattered. True happiness was a rare and fleeting sensation; she barely remembered the strength it could provide. Mélodie was tired of her dull and futureless life. She was lost and looking for purpose. I imagined her as a marginalized teen with a difficult and scarring past.

The honey-tongued Bilel might be the spark of hope she needed to restore her faith in life. The terrorist tried to discern Mélodie's jihadist motivations. He was like a salesman before making a pitch; he sought to understand the expectations and weaknesses of his prey. For him, Mélodie represented a type. Once he managed to categorize her, he simply had to churn out appropriate responses in deep and convincing tones. Bilel was an evil genie. He was an expert salesman, who was careful not to make direct queries on her plans for *hijrah*. Instead, he asked what she hoped to find once in Syria. It was an important nuance. Bilel still didn't know much about Mélodie—not her age, the color of her eyes, or her family situation. He wasn't concerned about any of that. In fact, he seemed interested in only one thing: that she had converted to Islam.

And for Bilel, Mélodie's faith was so strong that it would be easy to convince her to join him in the most dangerous country on earth. He only questioned Mélodie on her opinion of jihadists. I felt as if I were being polled. My answers relied on opinions I'd heard expressed in reports on high-risk suburbs.

"I've heard about what Israel is doing to Palestinian children. I've seen dozens of videos showing dead babies. I started following some of your brothers on Facebook who have left to do jihad, there and in Syria. Some mujahideen do good, others evil, so I don't know what to think."

"Focus on the good! I myself am an important mujahid. I've been devoted to religion for a long time, and I promise you: I can be very, very gentle with the people I love, and very, very hard on nonbelievers. I hope you're not one of them—"

"How could I be? I converted."

"Good, but that's not enough. It's not enough simply to pray five times a day and observe Ramadan. According to the Prophet, if you want to be a good Muslim, you must come to al-Sham* and serve God's cause."

"But I can't leave my family and abandon everything."

"Wrong answer! Let me guess: you're a capitalist."

Mélodie wasn't an intellectual. She wasn't interested in capitalism. Besides, what did it have to do with her family? She didn't understand what Bilel was implying. Soon he'd tell her to turn her back on the consumer society in which she'd grown up and embrace the Islamic court (or sharia law, a radical Islamic doctrine that exists in a minority of countries). Bilel was clear: Mélodie should not obey the

* *The Middle East.*

laws of her country. The only laws that applied to her were those of a radical form of Islam. A "pure" Islam, the one he'd embraced. Mélodie was naïve; she didn't see anything coming. She was duped at every turn. She didn't even notice the contradiction between Bilel's attack on consumer society and the fact that he was wearing the latest Ray-Bans and Nike apparel.

"Isn't capitalism about finding a balance between supply and demand? Something like that? LOL."

"Capitalism, my dear, is a blight on the world. While you're busy eating Snickers bars, watching MTV, buying Booba or Wu-Tang Clan albums, and window-shopping at Foot Locker, dozens of our people are dying every day so that we Muslims can live in our own state. While we're out risking our lives, you're spending your days doing meaningless activities. Being religious means imposing your values. I'm worried about you, Mélodie, because I sense that you have a good soul, and if you continue to live among *kafirs*, you'll burn in hell. Capitalism is exploitation of man by man, do you know what I mean?"

Now he was referencing Marx. Did he really grasp the German philosopher's doctrine and his concept of class struggle? Or was he simply repeating something he'd heard from someone else? I thought of Guitone, the Islamic State's "publicist," who dressed head to toe in Lacoste. Mélodie was stunned by the fate Bilel described for "*kafirs*." Her life in the West offered no hope, but was it really so bleak, compared to what Syrians endured? Bilel sought to infuse her faith with fear. He succeeded in sowing doubt in her mind and making her feel extremely guilty.

Abu Bilel was diabolical. I examined his profile picture. He was rather good-looking. The stunning grammatical errors barely distracted from the force of his conviction. What had drawn him to radicalism? What had made him so blindly committed—and therefore particularly dangerous? Some parents of jihadists compare the indoctrination of their children to methods used by cults. There was something of that here. Bilel acted as a kind of guru, who presented war as a divine mission. Mélodie was to accomplish her mission for the sake of a prophecy she didn't understand. I lit another cigarette.

"Are you saying that if I don't go to al-Sham, I'll be a bad Muslim, and I'll never know heaven?"

"Exactly . . . but you still have time. I'll help you. I'll be your protector. Can I ask you a question?"

Another smiley face; it had been a while. Mélodie had the choice between Syria and hell. Bilel painted a postcard of Syria that sounded pleasant and not at all hellish.

"I've checked out your profile," he went on, "and I only found one picture. Is it of you?"

Crap! I'd completely forgotten about the picture. I'd created Mélodie's account six years before, when the wives of extremists could still show their faces. Now the few Islamist radicals who allowed their wives access to social networks made them cover their faces. I hadn't thought to erase the old profile picture depicting a pretty, fair-haired girl.

"It's a picture of my older sister," I improvised. "She hasn't converted, so she doesn't cover her face, but I do."

"You scared me, *Masha'Allah*! Nobody should be

allowed to look at you! A respectable woman only shows herself to her husband. How old are you, Mélodie?"

Until that point, I'd felt like I was conversing with a car salesman; now I had the disturbing sensation of speaking with a pedophile. I wanted to tell him that I was a minor—just to see his reaction. But if I decided to meet him over Skype, that wouldn't work. I was just over thirty. And even if I looked young for my age, I couldn't pass for a pubescent teen.

"I'm almost twenty."

"Can I ask you another question?"

He clearly didn't care about Mélodie's age. What if she'd been fifteen? Would he speak to her differently?

It was midnight in Syria, eleven o'clock in France. My pack of Marlboros was empty. I was exhausted, and I sensed his next question would finish me off for the night.

"Do you have a boyfriend?"

Touché. I'd been dreading this moment. Mélodie would have to be succinct. She couldn't give any details.

"No, I don't. I don't feel comfortable talking about this with a man. It's *haram.** My mother will be home from work soon. I have to hide my Koran and go to bed."

"Soon you won't have to hide anything, *Insha'Allah*! I just want to know if I can be your boyfriend?"

"But you don't know me."

"So?"

"So what if you're not attracted to me?"

---

* *Forbidden.*

"You're sweet. It's your inner beauty that counts. I have a good feeling about you, and I want to help you lead the life awaiting you here. It breaks my heart to hear that you hide to pray. It's something I fight for every day here, to make others respect sharia."

His exploitation of Islam enraged me. Islam, and this is my opinion, is a great religion that encourages its believers to have sympathy for others. I'm agnostic, but I admire this community of people that finds its bearings throughout the world. André Malraux predicted that "the twenty-first century will be religious or will not be at all." This quote is often taken out of context; Malraux was referring to spirituality and "lofty" feelings. Bilel promoted a doctrine that, among other antiquated practices, forced women to wear full veils and marry at the age of fourteen. Some of these laws are intolerably violent: adulterous women are stoned to death, while men who cheat on their wives are merely fined; thieves pay for their crimes with their hands. ISIS seeks to install sharia law, first in the Middle East, then throughout the world.

When it came to sharia, Bilel was professorial: Mélodie was not to show an inch of her body, not even her hands, to anyone. A veil covering all but the oval of her face was not enough. She needed to wear a burqa and an additional veil over it. His pronouncements grated on me.

"My mother raised my sister and me by herself," I began, trying to calm things down. "She works two part-time jobs to make sure we have everything we need. I converted in secret. She isn't preventing me from practicing my religion."

"I'm sure your mother is a good person; she's just lost her

way. I hope she'll soon return to the right path—the one and only—Allah's path."

I was dumbstruck by his rigid thinking, bad faith, and blind judgments. Still, his arguments were relatively coherent, if ideologically impoverished. Bilel met Mélodie's questions with the most basic doublespeak: all the answers could be found in Islam—the medieval version of Islam promoted by ISIS. This conversation had gone on too long, and it was time to put an end to it. Mélodie reminded Bilel that she had to go to bed. He gave in and wished her sweet dreams, then added, "Before you go to sleep, answer me something: can I be your boyfriend?"

I logged off Facebook.

We'd exchanged one hundred twenty messages in the space of two hours. I carefully reread them all. Late in the night, I called Milan.

Monday

I woke up early, which is unusual for me. I rushed to the magazine where I often do freelance work, eager to discuss my weekend with one of the editors in chief. He also tracked the growth of Islamist extremist groups on the Net. Twenty-four hours earlier, I'd forwarded him the video of Bilel showing off the contents of his car. He was stunned by how easily contact had been established. He agreed that this was a unique opportunity. Information obtained in this investigation could provide a singular perspective on digital jihadism. However, he reminded me that pursuing this could be dangerous. Urging caution, he also assigned me a photographer, André, one of my best friends and also a freelancer. We'd worked together for years, and we made a good team. I would agree to Bilel's request to meet over Skype, and André would take pictures. After me, André would be the second witness to Bilel and Mélodie's relationship.

It suddenly struck me as strange to be playing one of two protagonists in a fabricated story, with both of us dealing in half-truths. I had never done something like this before, and it was troubling. So far, Bilel had been an evil genie I could consult whenever the need arose. Now I found myself implicated in the story. I would have to satisfy his need for domination, but for the time being, I was preoccupied with a single, urgent detail: how to become Mélodie. I needed to look ten years younger, find a veil, and somehow slip into the skin of a very young woman. Another editor, a former reporter who would also be supervising my investigation, lent me a *hijab** and a black dress—a kind of djellaba. Bilel wouldn't speak with Mélodie if she didn't hide the majority of her body. He was thirty-eight, and his attitude toward women was not the same as that of a young, newly recruited jihadist.

I was glad to wear the veil. The idea of a terrorist becoming familiar with my face didn't thrill me, especially not when the man in question could return to France, his home country, at any moment.

André arrived at my apartment that night around six o'clock. It was one hour later in Syria. That gave us about sixty minutes to prepare before Bilel "got home from fighting" and contacted Mélodie. We looked for the best angle from which to take pictures of the computer screen and keep me as indistinct as possible. We had strict orders to prioritize André's and my safety above all else. While André made adjustments in the living room, I pulled on Mélodie's somber clothing over my jeans and sweater. The floor-length black djellaba featured a small satin knot at the waist and was

* *The* hijab *is a veil that covers the hair and neck but leaves the face exposed. It can be compared to the chador. Veils that cover the face completely are called* niqabs, *burqas, or, in some countries,* sitars.

surprisingly fitted. I took a picture with my phone of the long train covering my Converse sneakers. I really looked like I was twenty years old. When I returned to the living room, André burst out laughing. "It's supposed to cover more of your forehead," he said, mocking me as he snapped a picture. He helped me readjust the *hijab*, which should cover every strand of hair and only show the oval of the face. I've worn burqas before, while working on other stories. I've never found them suffocating, as some women describe. People tend to regard you as oppressed when you wear a burqa, but the piece of clothing itself has never bothered me. The *hijab*, however, was a new experience. It reminded me of the horrible hoodlike knitted caps my parents forced on me as a child. It made my skin itch as it had when I was five years old, and my face looked like a puckered fish. André's hysterical laughter didn't help things.

I removed my rings, assuming Bilel wouldn't appreciate such frivolousness. Besides, if I wanted to become Mélodie, I had to remove all distinctive signs of myself. She wouldn't wear flashy rings. I also covered the small tattoo on my wrist with foundation. I had meant to buy nail polish remover to erase the bright red from my fingernails. I'd forgotten. Oh well. If Bilel said anything, I'd make up some excuse.

The hour was approaching. Perceiving my feelings of impatience, excitement, doubt, and fear, André tried to calm my nerves by talking about something else. To be clear, I wasn't afraid of the terrorist I was about to meet; I'd Skyped with others like him before. Rather, I sensed I was about to learn a lot, and I was afraid Mélodie wouldn't be able to handle it. As soon as I turned on my computer, I saw that

Abu Bilel was already logged on to Facebook and waiting for Mélodie.

“Are you there?” he asked impatiently.

“Are we meeting on Skype?”

“Mélodie?”

“Hello? LOL.”

“Mélodie???” . . . “Sorry: *salaam alaikum* . . . ☺ You there???”

Monday, 8 p.m.

Okay. It was time. I sat cross-legged on my sofa. It had a high back, which hid most of my apartment—and any distinctive features—from the camera. André had also removed from the wall a famous photograph taken in Libya three years earlier. He positioned himself in a blind spot behind the sofa. Mélodie bought some time by typing a reply to Bilel. My smartphone was already recording. I was also equipped with another, prepaid phone I'd bought a few hours earlier. The Islamic State is brimming with counterespionage experts and hackers. It was safer if Bilel didn't know my phone number, so Mélodie had her own. I'd also created a new Skype account in her name. I'd found a video on YouTube explaining how to scramble an IP address. If things started to go wrong, Bilel wouldn't know where to find me.

The Skype ringtone sounded like a church bell tolling in a dreary village. If I pressed the green icon, I would become Mélodie. I took a moment to breathe. Then I clicked the button, and there he was. He saw me, too. For a split second, we didn't speak. Bilel stared at Mélodie. His eyes were still accentuated with dark liner. They smoldered as he gazed at the young Mélodie, as if trying to cast a spell. I don't know if it was because I was nervous about meeting this man face-to-face, but in any case what captured my attention most was his location. Bilel was Skyping Mélodie from his car, using a state-of-the-art smartphone. He lived in a country often deprived of water and electricity, yet he had access to the latest technological devices. The connection was good, which was not always the case in such circumstances. Bilel made ISIS sound more like a nongovernmental organization than a terrorist group, but one would never confuse him with a humanitarian aid worker. He looked clean and even well-groomed after his day on the front. He was a proud man, his shoulders pulled back and his chin thrust forward, but I sensed he was nervous meeting Mélodie. After what felt like an eternity, he finally broke the silence:

"*Salaam alaikum*, my sister."

I made my voice as tiny, and as sweet and bright, as I could, considering I'd smoked like a chimney for the past fifteen years. And I smiled. My smile instantly became my best defense mechanism, and it remained so throughout my investigation. I would use it whenever I didn't know how to react. I believed I could become another woman by playing the understanding friend. But I couldn't bear the thought of watching the videos André was going to film of these virtual discussions. Today, when I watch them, I don't see the pure and naïve Mélodie; for me, she isn't the person I see smiling and impressed as she converses with Bilel. I see myself, Anna, dressed in black, sitting on my familiar couch, which I have come to hate. I'm the girl smiling. It isn't Mélodie; she doesn't exist. Should I feel ashamed for having taken part in this exercise? I'm a private person, and when I see these images of myself—playing a part, but it's still me—I feel sick.

Mélodie replied using the same polite expression, but she didn't finish her phrase. André distracted me by jumping around the sofa and waving his arms, careful not to enter the camera's field of vision. In the heat of the action, I hadn't replied correctly to Bilel's question. The proper response to "*salaam alaikum*" is "*walaikum salaam.*" It was a beginner's mistake, and I knew better. I wanted to laugh, and at the same time, I would've liked to see André in my shoes! But I couldn't do anything; Bilel was hanging on Mélodie's every word. He may have been in Syria, and I in France, but our faces were separated by mere millimeters. I had to be careful not to let my eyes wander from the screen. I was flooded with random thoughts. Ignoring André, who was still jumping around like a kangaroo, I choked when I heard Bilel's first question.

"What's new?"

Seriously? I hadn't expected him to show interest in Mélodie's day. I was so caught off guard by this ordinary request that I couldn't think of anything to say but, "So much! But I'm shy. First, tell me about yourself."

"What do you want to know?" he asked, smiling confidently.

He took the bait. Mélodie's life didn't seem to interest him much after all. Too bad for her. Great for me. That said, I didn't want to awaken his suspicions and risk blowing my cover by asking too many questions. ISIS knows that many journalists and police officers hide behind fake identities. Mélodie was twenty years old, and her knowledge of the world needed to match her age. She didn't know much about politics, geopolitics, or holy wars.

"It's crazy to be talking to a mujahid in Syria," she said, impressed. "It's like you have easier access to the Internet than I do in Toulouse! I share the computer with my sister, and my mom takes it away from us a lot. And you're totally in a car. It's insane! Even your phone is newer than mine."

In addition to getting into character, I was giving Mélodie a plausible excuse for future unavailability. She lived with her family, and she couldn't always honor her engagements.

"Syria is amazing. We have everything here. *Masha'Allah*, you have to believe me: it's paradise! A lot of women fantasize about us; we're Allah's warriors."

"But every day people die in your paradise. . . ."

"That's true, and every day I fight to stop the killing. Here the enemy is the devil. You have no idea. The enemy steals from and kills poor Syrians. He rapes women, too. He's attacking us, and we're defending peace."

"Is the enemy the president of Syria?"

"Among others. We have many adversaries."

In addition to Bashar al-Assad's regime, he mentioned the al-Nusra Front (an armed branch of al-Qaeda), Syrians, and all those he considered infidels. I knew ISIS wouldn't hesitate to decimate the Syrian people (already oppressed by the Alawite dictatorship) if they refused to adhere to the terrorist organization's rules. But I sensed the fighter didn't want to elaborate. Bloody descriptions of the violent acts he committed every day didn't fit with his strategy of lobotomizing his prey. Especially not when they might involve the weak. He didn't want to give Mélodie pause.

"You're awfully curious," Bilel said. "Tell me, do you wear your *hijab* every day?"

Mélodie recited what I'd heard from the majority of girls I'd met during my career who had secretly converted to Islam.

"I dress normally in the morning. I say goodbye to my mom, and when I'm outside the house, I put on my djellaba and my veil."

"Good. I'm proud of you. What you're doing is really brave. You have a beautiful soul. And you're very pretty on the outside, too."

Bilel peered lecherously at Mélodie. She asked him to show her his surroundings. He claimed to be near Aleppo. In reality, he was probably several miles outside of Raqqa—ISIS headquarters and the first city where the organization literally established a state whose laws and strict policies subjugate locals through barbaric practices.

"The Prophet says you must choose a wife based on her character. Her inner beauty is her true beauty," he added. "But when a woman is beautiful inside and out . . ."

Bilel bit his lip and stared at what he could see of me. I smiled. In response to Mélodie's request, he got out of his car and his smartphone showed me images of a devastated Syria. Not a person in sight. It was about 9 p.m. in Syria, and it was absolutely silent. Suddenly, men's thick voices broke the mournful silence.

"Don't say anything!" Bilel ordered anxiously. "I don't

want anyone to see or hear you! You're my jewel; you're pure. Okay? Do you understand? Tell me you understand."

Mélodie said she understood. She wouldn't make another sound until he instructed her otherwise. That meant I could listen to the conversation. I was able to distinguish the voices of two other men. They greeted one another in Arabic, then switched to French, which sounded like their mother tongue. They laughed a lot, congratulating themselves for having "slaughtered them."

"*Salaam alaikum*. What's up?" one man asked. "Are you putting in overtime, or something?"

"I'm on the lookout, brother, lookout duty . . . nothing special. Nothing happening here. This area is all cleared out. You know that."

As he finished talking, a sardonic smile spread across his face. Given the camera angle, I was able, with difficulty, to make out his facial expressions. By "cleared out," Bilel meant his militia had laid siege to the area. The dried blood I saw on the concrete was evidence of the attack. ISIS's black flags with white insignia floated in the distance. I listened to Bilel go on about a variety of issues, notably his impatience for the arrival of his "American cargo" and "chocolate bars." André and I exchanged a meaningful look. The other men seemed to treat Bilel with respect, and they were quick to congratulate him. The exchange was too short to draw any conclusions, but their way of politely addressing him suggested my "contact" was higher in the ranks than they were. A minute later, he said goodbye to his fellow fighters and spoke into the phone, worried Mélodie might have hung up.

"Oh, you're still there! And just as beautiful—"

"Who were they?"

"Fighters who came to say hello."

"Oh, it sounded like they were reporting to you. You don't want to brag, but I bet you're a boss or something."

"You're right; I don't like to brag . . . but people respect me."

"Why? Are you an emir?"

Bilel adopted an attitude of false humility.

"You've guessed it, but I don't like to brag. Let's keep it between us. We're all here for the same thing."

"You seem really determined . . . can I ask what your job is?"

"Killing people."

"Killing people is your job? I mean, that's a job?"

"Of course it is! I work hard here. This isn't Club Med!"

"You kill infidels?"

"Yeah, and traitors, too. I kill anyone who tries to prevent Islam from dominating the world."

"What do you mean? Do you plan to take over the world?"

"Our leader Abu Bakr al-Baghdadi seeks to abolish all borders. It will take some time, but soon the world will be one big Muslim territory."

"And what if the world is against it?"

"Then I'll have a lot of work to do. In time, we'll succeed."

"A lot of work to do? Will you kill everyone who doesn't agree?"

"Me and my men will. I can't do it alone! *Masha'Allah.*"

"I bet you helped capture Raqqa. There were pictures of the Islamic State everywhere."

The Battle of Raqqa, which took place in March 2013,

was one of ISIS's bloodiest victories. It demonstrated to the world the group's fearsome determination. Fighters waved the black standard throughout the city and posted enemy heads on spikes in a main square. Pictures of mutilated bodies spanned the globe, acting as weapons of propaganda. Even Mélodie had seen them, on Twitter. Needing to remain focused, I shifted into autopilot. There would be time to think about Bilel's madness later.

"You make me laugh!" he said. "Yes, of course. We obliterated them. It was crazy. . . . I'll send you some pictures."

He really did send them. This grisly memory made him extremely happy, and he didn't try to hide it.

"Anyway, you're not interested in all that," he went on. "You ask too many questions. Tell me about you!"

"I want to know one thing first. . . . You say you kill bad people to cleanse the world. But why do you mutilate them? If your cause is noble, why such barbarism?"

"Well, we* conquer territory by eliminating people. But everyone has a specific job. I don't mean to brag, but I'm very important, so I'm in charge of supervising operations. I give orders, and when all the *kafir*s are dead, the emir decides what to do with their bodies."

"What does that mean?"

"Well . . . didn't you say you'd seen the videos and pictures? That day, for instance, the emir of Raqqa told us to cut off their heads. But enough of that. Tell me about you!"

---

* *ISIS's brigades.*

"Okay, but I'm too shy! Let me see your car first. It looks like you have a lot of interesting stuff."

Bilel was glad to show off his car, delighted whenever Mélodie—whom he already considered his betrothed—flattered him. Mélodie told him she thought the white submachine gun sitting amid a heap of clutter on the backseat was pretty. Bilel grabbed it and offered to give it to her. Laughing, he said, "I'm not surprised you like it! Women love this model because it's easy to use. Do you like guns? I'll give you plenty, starting with a lovely Kalashnikov."

I could tell from his expression that he was being sincere.

"I want to learn more, but what does all this have to do with religion?"

"What guided you to Allah's path?"

I was dying for a cigarette. At that moment I couldn't think of anything else. Mélodie had existed for years without really existing. She'd simply been a name on a Facebook profile. As late as that morning, I never would have imagined playing this part for Bilel or needing to fabricate a backstory for a desperate and ultrasensitive young woman. I hadn't had time to invent a "real" life for Mélodie. My veil was starting to itch, and when I glanced at André, a man known for his hyperactivity, I noticed that he was stunned.

At a loss for words, Mélodie stammered, "My dad left when I was little, and whenever my mom was too overwhelmed to take care of us, we stayed with my uncles. One of my cousins was Muslim, and I was fascinated by the inner peace that his religion gave him. He guided me to Islam."

"Does he know that you want to come to al-Sham?"

Bilel assumed that everything had been decided. For him, Mélodie would soon arrive in Syria.

"I'm not sure that I want to go—"

"Listen, Mélodie. Among other things, it's my job to recruit people, and I'm really good at my job. You can trust me. You'll be really well taken care of here. You'll be important. And if you agree to marry me, I'll treat you like a queen."

Monday, 9:30 p.m.

Marry him?! I logged off Skype as a kind of survival reflex. Pulling the *hijab* down to my neck, I turned toward André, who looked as dumbfounded as I was. We stared at each other, incapable of saying anything other than, "Oh, shit!" We knew we could put a stop to everything right then; that night could become another anecdote we'd tell. But, of course, we wouldn't stop now. We wanted more. That was the goal of any investigation: knowing more. If Bilel had proposed in person, I would have run, but there was a screen separating us. That was an important distinction.

"What an a-hole!" André exploded. "He wants to marry you, now?" he screamed at me as though speaking directly to Bilel.

André was familiar with the Islamic State's methods of propaganda, but suddenly he was faced with an unspeakable truth. He was the father of thirteen-year-old twins, and the idea of the terrorist organization targeting children sickened him. André was born in France. His father was one of the Kabyle, a people from eastern Algeria, and his mother was from Spain. He believed in God, but aside from lighting a candle at church whenever he had a very pressing wish, he didn't practice any particular religion. He simply had faith. In his youth, he'd known hotshot men like Bilel. Back then, the French state turned a blind eye to petty crime. André had a visceral loathing for the leadership ISIS had installed by force. How was I to respond to Bilel's proposal? He suggested explaining that since Mélodie wasn't married, she didn't want to arrive in Syria alone. If she decided to go at all.

Bilel called back. André held out a cigarette and I took a drag. The use of tobacco and of alcohol is strictly prohibited and severely punished by ISIS. When she answered, Mélodie blamed the interruption on a bad Internet connection and immediately launched into the explanation André had provided. She added that if she decided to go to Syria, her cousin would accompany her. First, because respectable women didn't travel alone. Second, because her cousin wanted to help the cause.

"If you want, but I don't see why," Bilel said, upset. "You don't need him. Dozens of girls arrive by themselves every week. You're not as brave as I'd thought, Mélodie."

Twenty-year-olds are obsessed with showing off their bravery and commitment, and that's what Mélodie did.

"You don't think I'm brave? You obviously don't know me very well. If I have to leave everything to do my jihad, I want answers to my questions, and I want to travel with my cousin. If I'm going to fight, I want to know why."

"Oh yeah? And what's your life? If your cousin were a true believer, you would know. . . . But if you really want him to come with you, fine, do what you want."

Bilel was visibly annoyed. I didn't realize it at the time, but mujahideen say that "guiding people to Allah's path" offers a guaranteed pass to paradise.

"Do you not trust my cousin? Or did you want me to come alone?"

"Do what you want, but don't you have girlfriends interested in *hijrah*?"

There it was. I couldn't wait to see how he'd try to convince Mélodie to bring along a cargo of prepubescent girlfriends. André was unable to suppress an angry sigh.

"I don't know. I haven't told many people about my religion. What difference would it make if I came with a man or a woman?"

"It wouldn't. Only, women in Europe are treated badly and used like objects." He sighed. "Men show you off like trophies. I want lots of people to join ISIS, but I'm especially interested in recruiting those who are treated the worst, like women."

He didn't give me time to react.

"Mélodie, answer me. Do you want to be my wife? Mélodie, did you hear me? Do you want to marry me?"

"I . . . I mean, that's much too important and personal to discuss here, and so soon."

It was strange to have to simper and act coy with this madman in front of André. He thought of me as a little sister, and he knew my boyfriend. I disabled the video connection. Bilel could continue his conversation with Mélodie, but he wouldn't be able to see her.* I did that for myself. It

---

** Skype conversations can be audio and video, or simply audio.*

felt like his face had invaded every corner of the room, and I didn't want to see it anymore.

"My friend Yasmine is Muslim," I said, changing the subject, "and she complains about not being able to practice her religion in Toulouse. I could invite her to come with me, but I'm not sure if she's allowed, since she's a minor."

"Of course she can come!"

"She's only fifteen."

"I fight for sharia law every day. Here, women are supposed to get married when they turn fourteen. If Yasmine comes, I'll find her a good man, someone who will take good care of her. Matchmaking for European women is a job here. The women wait in a hotel and we introduce them to single mujahideen."

Yasmine didn't exist, but I wondered how many real Yasmines were being lured at that very moment by men like Bilel.

"Bilel, I have to hang up. My mom is getting home."

"I'll be here tomorrow after the fighting, as usual, at seven. *Insha'Allah* . . . Good night, my baby."

My baby?

I logged off. It was stuffy. André opened the window. We were surprised, not by the content of the exchange, but by how rapidly everything had unfolded. While I paced the room, André cursed Bilel. I felt consternation, anger, indignation, but also some satisfaction over our first foray into the mind of a killer. It had been a success, but it wasn't going to be easy. I would have to put up with his cruel ideology and play along, but he was also a pawn in Mélodie's game. At no point during our conversation had he seemed suspicious. I'd gained his trust, and we sensed that could lead to something important. Would the benefits outweigh the costs? Neither of us knew, and in search of an answer we pored through everything the terrorist had said. I'd removed my veil as soon as I'd signed off Skype, but I was still wearing the black dress. When I stood, I tripped on its train. André, who usually seizes any opportunity to tease me, barely noticed. When he left my apartment, he was disconcerted and high on adrenaline. He bombarded me with worried text messages until late in the night, reiterating the enormous risks inherent in this investigation. Although we may have let ourselves make fun of Bilel, we both knew that the man behind the screen was extremely dangerous. André wanted to continue our investigation, but he thought it best not to aggravate the jihadist. Retaliation could be devastating. "Let's make it short and sweet, Anna," he said, "and then move on."

Mélodie

Mélodie's difficult life made her into a ticking bomb. She didn't wish anyone harm, except perhaps herself. Life itself was flaying her alive. She'd been mourning her father's departure for the past twenty years, even if she'd never been able to call him dad. She blamed herself for his absence. He'd never wanted to marry, much less have kids. She was his second child, and her birth had been the last straw. He never recognized Mélodie as his own. Ever since she'd found out, Mélodie had been compensating for her broken family. She comforted her friends when they were overwhelmed and mended their broken hearts. She was the friend others turned to when they needed an ear. Helping others made her forget her own feelings of helplessness. It only took a moment, and she felt less empty. Still, she lacked self-confidence, and although she'd been searching for meaning her whole life, she'd never really thought about the future.

If Mélodie could have expressed her pain in words, she would have told her mother that, in the end, perhaps their family of three was better off without a father. After all, her brave mother had done a fairly good job of raising her daughters by herself. Generally, children who become jihadists keep in touch with their mothers. Although they're taught to forget their past, these children rarely sever the maternal bond, which remains their one point of contact with their origins. Like all terrorists, Mélodie was moody and unpredictable. Her mother hustled for work, anxiously trying to make ends meet. Lacking structure, Mélodie got her education from the streets. When she was younger, Mélodie imagined her mother having to identify her at the morgue. She loved her mom, but aside from disillusion and fear, she didn't have anything in common with her. Unspoken wounds made her tough and rebellious. Then, as the silence endured, she became empty. Empty of love, empty of hope.

Back then, she used to hang out with a group of girls known in the neighborhood for shoplifting and getting into fights. She wasn't overly enthusiastic about these girls, but they were a comfort on those long days when she skipped school. The girls—all minors and armed with braces—spent countless nights looking out for cops while others shoplifted. They liked to brawl, with anyone—even guys—over anything. But they spent most of their time hanging out in parking lots or parks, drinking Fanta, eating Filet-O-Fish sandwiches, and dancing to whatever music was currently popular. Mélodie was often bored. She didn't think she was interesting and was surprised to belong to a group. She couldn't bring herself to care about their discussions of the latest reality show or some girl losing her virginity. Because she felt different from the others, she figured there must be something wrong with her. And although her personal history was sad, it was nothing compared to the stories she heard from the other girls. Whenever her friends confided their problems, Mélodie comforted them, but she never spoke of her own unhappiness. She didn't like drawing attention to herself. She didn't want people to feel sorry for her. She just wanted to be loved.

Over time, she grew indifferent to the world around her, but the girls protected her. They all came from somewhere else, from immigrant families. Mélodie was white and only knew her father's first name. She always wished she'd been born elsewhere, but she didn't know exactly where. She smoked pot for a while, then grew out of it. Marijuana was yet another source of disillusionment. In the neighborhood where she'd lived her whole life, she often heard rumors—usually false. Too many rumors. Sometimes she'd listened, blindly testing taboos and dabbling in petty crime. The line between right and wrong became less stable and more porous. But as with other experiences, after the adrenaline wore off, transgression left Mélodie feeling miserable. Ashamed, she could never look her mother in the eye after a day spent at the local precinct. Back then she'd been trying to prove something to herself, to fill an internal void. Her credulous nature made it difficult to differentiate between good and bad behavior. In fairy-tale terms, she would have chosen to be Robin Hood over Cinderella. Growing up in a tough neighborhood, with a brokenhearted mother, she'd developed a survival instinct that protected her from some vices.

She'd never been very interested in boys. A few unmemorable boyfriends quickly satiated her appetite for love. She was waiting for a real love story, the kind that drove you mad, the kind her mother had gone on about ever since Mélodie was a little girl. Mélodie was unconsciously looking for a father more than a lover. A man who could protect her, a man with the strength to give her a sense of purpose in life. A person she could trust absolutely. Someone mature, someone like Bilel. Her life was a desert and he was an oasis. He was the antidote to a lonely and lackluster life like her mother's. Mélodie saw in him a solution to her troubles. Coming to the aid of the Syrian people seemed a much more ambitious life than the one she imagined for herself in France. The deadening pain inside her was indescribable. She wanted to scream and cry. But she'd learned early on that expressing one's troubles was a sign of weakness. And weakness was despised where she came from.

Mélodie spent hours cooped up in the room she shared with her older sister, who was rarely there. On the walls, posters of Scarface abutted pictures of Rihanna and the rapper Mister You. She liked being by herself and listening to the radio with the volume turned up. Shut inside her music box, she felt the weight of despair lift from her shoulders. She listened to the French rapper Diam's and checked out blogs and Instagram accounts, losing all sense of time. Diam's also grew up without a dad, and abandonment is a common theme in her songs. I listened to her albums to get a better understanding of Mélodie. In fact, the artist's lyrics described the character I was trying to create. Here was a tormented girl rapping about her trials and feelings of loneliness.

Many people under the age of twenty-one experience the kind of desperation Diam's talks about in her songs. ISIS offers them a way to fill the void in their lives. The terrorist organization targets all social milieus, enticing young people with promises of a purposeful life. ISIS accommodates everyone, from those eager to fight to humanitarian aid workers. The organization propagates the illusion that it cares about these kids so that it can dispossess them of their past and reprogram them. It's like a cult leader luring the faithful. Its favorite weapon is the Internet, which it uses to transform would-be jihadists from cyberpawns into cannon fodder. The proof: in less than forty-eight hours, Mélodie was promised true love, marriage, and an idyllic life.

But for some reason she was hesitating. Leaving her family terrified her. Despite fights, which can be expected between a single mother and her daughters, Mélodie's family had always been loving and supportive. So Mélodie tried to convince herself by watching dozens of videos on YouTube. As far back as she could remember, she'd heard that Americans were monsters who tortured Muslim prisoners in Guantánamo. Her heart went out to children suffering in Palestine and Syria. She blamed their fates on the West. Everybody in her neighborhood said that the French state and the Jewish community had conspired to create Mohammed Merah, also from Toulouse. They had orchestrated the barbaric killings. The man on the scooter had simply been the scapegoat in a plot to tarnish the reputation of Muslims in France. She didn't know what to think of that rumor. Killing a child went against the teachings of Islam. But Bilel had called the killer a servant of Allah.

She watched video after video, including one that featured Omar Omsen, a thirty-seven-year-old French-Senegalese man who was wanted by both the French and Belgian police. He was thought to be one of the brains behind an important jihadist recruitment network that sent people from Europe to the Middle East. In October 2013, the different branches of the DCPJ (Direction Centrale de la Police Judiciaire)* were furious when they discovered the successful departure of seven members of the same family in Nice, including four children. Omar Omsen had organized their journey. He often posts videos on YouTube of himself praising sharia law, encouraging viewers to ignore the laws of their countries and uphold only Islamic—sharia—law. He tries to brainwash and shame them, saying things like "A good Muslim doesn't live in a country of nonbelievers. If you aren't helping the Islamic State, then you're murderers. While you say your prayers and pore over the Koran, others are fighting for the one and only Allah, who wants us to establish a worldwide caliphate." Most jihadist recruiters these days don't just hang out around mosques in djellabas. Like them, Omar Omsen lives in a European country, safe from bombings, and with all the amenities he could desire.

Mélodie watched another video of him on a boat. His blue eyes shone as he compared the white froth of the waves to the purity and sense of plenitude that Islam could offer if the viewer rigorously applied him or herself. Mélodie let her imagination wander, mesmerized by the immaculate white foam that was supposed to represent her religion. Bit by bit, she gave in to the thrall of these seductive stories and

---

** Translator's note: The DCPJ is the French national judicial police. It investigates and fights serious crime throughout France, with the exception of greater Paris.*

promises of meaningful encounters. People say that no man is an island. The brilliant antiterrorist judge Marc Trévidic has been explicit on this point. According to him, even though there exist some cases of jihadists working in isolation, they rarely decide to act by themselves. There is always someone close to them who has molded and encouraged them. To date, the Merah case still raises many questions, but in my opinion, he was influenced by his older sister, Souad, who acted as a mentor and guided him toward radicalism. Souad recently left Toulouse for Syria with her four children, the youngest of whom is one year old and named Mohammed, "in homage" to her "hero," a man she has publicly and privately praised. French authorities only learned of her departure once she was already in Syria. In Mélodie's case, Bilel would be her guide.

Mélodie's thoughts were scattered. Leaving Toulouse would mean giving up her boring commute from the Reynerie station on the A Line. She'd taken that train every day through the Mirail neighborhood for the past ten years, the music on her MP3 player her only distraction. She was tired of its worn-out seats and the mundane urban landscape. It had become a symbol of her dull existence. Life in Syria couldn't be worse than what she knew here; at least she'd know why she was getting up in the morning. She wondered what Bilel was doing at that precise moment.

Her survival instinct, gleaned from a lifetime in her rough neighborhood, told her not to fall for his charms. But it was too late. Mélodie already saw him as a king. And she had always dreamed of being a queen.

Thursday

As I did every morning that week, I awoke to find several affectionate messages from Bilel. I received more from him than from Milan. They all began with "My baby." It made me want to go straight back to bed. *Oggy and the Cockroaches*, a cartoon I like, was on TV. It provided a good transition before I went back to playing Mélodie. I listened to the radio: more news about a French minor leaving for jihad in Syria. I turned it off. The twenty-four-hour news channels were also abuzz over the "jihadist of the week." I turned off the TV and got to work on reading Bilel's missives. The fighter said he was leaving for battle, and he hoped Mélodie would have a good day. He said nothing about religion. He could have been any boyfriend ardently missing his new girlfriend. I wanted him to tell me about his army's plans, but he was more interested in flirting, which made me uncomfortable. I'd have to strike the right emotional balance and carefully mete out Mélodie's feelings.

Abu Bilel was a full-time job. During the day, I fact-checked his claims at the office. At night, my avatar took over, conversing with him over Skype and coaxing out new information. The day before, Abu Bilel had claimed again to be near Aleppo. Mélodie might have believed him, but thanks to Internet sites specializing on the Middle East, I was able to track the latest battles and land seizures. The Islamic State had abandoned Aleppo six months before. Aleppo, the second-largest city in Syria, was divided between loyalists and rebels, whom Bashar al-Assad's army regularly bombed. It seemed unlikely that Bilel was there. As I'd suspected during my first Skype conversation with Bilel, he was probably located in Raqqa, the Islamic State's capital.

When André and I met that night, we were both strangely calm. Were we getting used to these Skype sessions? Still, our anxiety grew as Bilel's call neared. The mujahideen always spoke with such zeal about atrocities, and it was difficult to sleep after these conversations. I also felt uncomfortable playing Mélodie's role and letting Bilel fawn on and manipulate her. I couldn't avoid his advances if I wanted him to trust me. I had to compliment him from time to time and smile like a woman in love. I had to play the part. It was the only way I could obtain a narrative of his days spent "cutting off heads." Only, I'm not an actress, and André's presence made this dizzying exercise all the more difficult.

As we finished making sure everything was in place, Bilel sent Mélodie a message asking her to call him. It had been preceded by several others:

"Mélodie."

"Mélodie??"

"Mélodie, my baby?"

"Mélodie???"

I logged on. Today he was alone in an Internet café. His hair was slicked back with gel, and he'd traded his fatigues for casual wear. His outsize self-confidence was still there. I attacked in an innocent voice, drawing on the concrete facts I'd learned earlier in the day.

"Are you okay? I was worried about you. My friends told me there was a very bloody battle today involving ISIS. Is that true? Where was it?"

"You're worried about me? That must mean you care about me—"

"Seriously, this isn't a joke. Answer me. Where was it? Did anybody die?"

The more Mélodie acted concerned, the more Bilel was flattered. The false look of humility on his face made his smile more arrogant than ever.

"You know I don't like to brag, but don't worry; Allah protected us again from the devil. Some rebels laid an ambush for us about twenty miles from where I am right now. They were trying to weaken ISIS's forces. But we're the best at what we do, and we're always one step ahead of everybody else. We killed them, and I can promise you, they won't be going to paradise."

"Did *you* kill them?"

"You ask too many questions! Let's just say, I slit a few throats. Anyway, I can tell you it was a bad fifteen minutes for them."

I knew Bilel was lying. How could he have spent the day chopping off heads and still have found time to leave more than a dozen voice messages for Mélodie? Besides, hadn't he told her in a previous exchange that he purposely stood on the sidelines so as not to put himself in danger? He was boasting to impress her. Still, it didn't matter if he'd slit men's throats today or yesterday; the timing didn't change his horrific actions. He'd been a cold-blooded killer for years, spilling blood in the name of a religion. The night before, he had told Mélodie that if he were in a Mexican cartel whose members represented each kill with a tattoo, he would be completely covered in ink. I imagined tattoos all over his arms, then remembered that tattoos were not allowed by Islam.

"Wow . . . you sure take a lot of risks," Mélodie commented, before continuing her morbid line of questioning. "How many enemies died, and what did you do with their bodies?"

"We killed at least twenty. Their bodies can rot in a mass grave. That's more than they deserve, but I'm not in charge of logistics. Don't worry about me . . . tell me about you, baby."

"I watched a lot of videos on the Islamic State this afternoon. Actually, I want you to tell me about it, since there are a lot of rumors."

"There's only one thing you need to know: the true Islam is the establishment of a caliphate. True Muslims are those who devote their lives to that cause, and that's ISIS. Everyone else is a nonbeliever."

"What kind of nonbelievers were you fighting today?"

"Those *kafir*s from al-Nusra. Don't worry, we let them have it."

Bilel smiled and held up his phone, showing me a picture of mutilated bodies. He was gloating.

"I didn't see it very well. Show me again."

"No, I'll save the best for when you get here."

"But were those severed heads?"

He smiled and winked.

"You kill people. . . . That doesn't fit with the Islam I know."

"Sister, war always precedes peace. And I want peace, as Allah has commanded. That way we can start a family here, together. *Masha'Allah*, my baby. You've never told me if you thought I was good-looking. Be honest."

From the very beginning of their exchanges, Mélodie had been avoiding this subject. I didn't know what to say. I couldn't turn back now. He spoke to me of marriage with increasing insistence every day. I was backed into a corner: I had to pretend to have feelings for a murderer. I had to flirt. And it had to seem sincere. Now more than ever I had to be a good actress.

"You're hot . . . and brave, which I admire in a man."

"Thanks. What else?"

"You have beautiful eyes."

I was trying to get by with minimal flattery. It already seemed like too much.

"That's a girlie compliment! Do you want to go further with me?"

"I don't know how to answer that. . . . You know better than anyone that a respectable woman doesn't speak to a man who isn't part of her inner circle."

"That's true, but I asked you to marry me."

"We'll talk about that, but give me some time to think. Anyway, you never answered my question. Where did the battle take place? Are you hurt?"

"You're so cute, and innocent! No, I'm not hurt. I'm the real deal. . . . If somebody wants to take me out, he'll have to get up early. You're pure, Mélodie, so I'm gentle with you, but I'm a nonbeliever's worst nightmare. Otherwise, most of the time I help with reconstruction work."

"Reconstruction? Where?"

"In a city near Iraq that was destroyed by the Syrian army. We have a lot of work to do there. I'll put it this way: we want to make the poorest city in Syria into the richest. And that's where we'll live, together and happy, *Masha'Allah*."

Bilel was referring to Deir ez-Zor, a city in eastern Syria on the banks of the Euphrates River, about three hundred miles from Damascus and near the Iraqi border. Not long ago, half the city was in the hands of Syrian rebels and the other half was controlled by Assad's regime. Using its usual bloody methods, the Islamic State had recently chased the rebels from the city and taken over the entire province, along with most of its oil fields. The Islamic State's religious jihad is also a jihad for oil. In fact, ISIS produces more oil than the Syrian government. The exact numbers vary depending on which political or religious group is providing them, but overall, it's estimated that the Islamic State produces $3 million of oil per day between Iraq and Syria. The oil production of Assad's government, meanwhile, has fallen to 17,000 barrels a day. With the Islamic State making millions of dollars a week, its army growing every day, and with its stash of heavy artillery, ISIS has become a formidable force. It won't be easy to defeat. Additionally, ISIS has plans to expand over more territories, as it did by moving from Iraq into

Syria. Potential territories include Libya, Jordan, and part of Lebanon. But again, I'm not an expert, and both Lebanon and the Jordanian monarchy have precious allies who won't let ISIS invade them. The day before, I was speaking with a mujahid over Twitter when I came across a picture of a little girl writing this on a wall in Raqqa: "Your jihad is about oil." Of course Bilel was careful not to breathe a word about this profitable business to Mélodie. The oil market was like capitalism: it wasn't that interesting.

"What are you doing to enrich the city? Are you building schools and hospitals? Oil is worth a ton of money. Maybe it could help with the reconstruction? It would bring in a lot of cash."

This time Bilel was the one caught off guard. He scratched his head and avoided Mélodie's gaze. He lowered his eyes and fumbled for words, searching for the right lies to quiet her doubts. He'd massacred countless men under the guise of humanitarianism, and it would be awkward to admit that it was also for money. A lot of money. I still relish these small moments when Mélodie innocently undermines him. At last he replied, his voice evasive but his expression unwavering.

"Yes, we have plans for that, among other things. But constructing hospitals is expensive and right now we have to build up the city's coffers. A lot of oil has been stolen from Deir ez-Zor these past few years, much of it by the current government. We in ISIS have therefore been working to regain control over oil production and make it profitable for the city. But it's a lot of work, and for the moment it isn't making any money. In a way, we're planting seeds and waiting for them to grow. But you shouldn't worry about any of that. Don't waste your time with questions about capitalism; ask me something else!"

My questions clearly bothered him. Mélodie had to regain his confidence. I'd have to put the topic of jihadism and oil on hold.

"Tell me about yourself and the life I'll have if I decide to come to Syria."

"Of course you're coming. You'll see, you'll make a whole new life, a very happy one, for yourself. You seemed to like my guns the other day. When you get here, you'll take shooting lessons for a couple of weeks, depending on your level."

"For self-defense or so I can kill nonbelievers?"

"It depends. You're allowed to kill, so long as it's someone who doesn't respect Allah. There's nothing wrong with that. Quite the opposite, actually. Married women have the right to accompany their husbands to the front. Sometimes we let our wives shoot—they love it! Generally, they like to film our exchanges with enemies."

"You mean I have the right to take someone's life if I believe that person isn't living according to Islamic law?"

"That's right. *Kafir*s are *haram*, and we can do whatever we want to them. You can burn them or strangle them, for example. Giving them a painful death is a service to Allah. *Insha'Allah*."

This distorted vision of Islam would have driven me crazy if I were Muslim. The horrors coming out of this honey-tongued monster's mouth didn't surprise me, but I was still disgusted. I owed it to myself to prod him a little on the issue.

"In the Koran it says that we can condemn those who don't respect our laws, but I don't remember it saying that killing men was a service to God."

"Since they're trying to eradicate us, it is. We represent God's will."

Mélodie was about to ask another question, but Bilel cut her off. He had a postcard to sell her. An ideal life.

"We're talking too much about the dead. It's beautiful here. There are so many things to see. The sea is magnificent, and the landscape is fascinating. You'll make a lot of friends here. You'll have a little group of girlfriends, and you'll do girl stuff together." He laughed. "It's a whole other world here . . . soon to be your world. During the day while I'm off fighting, you'll spend your mornings perfecting your Arabic and your afternoons doing whatever you want. You can hang out with your girlfriends or visit children in hospitals and orphanages."

"I can go out with my girlfriends, even if there's no man to accompany us?"

"You can if you behave appropriately. Besides, you European converts are insane. As soon as you arrive, you want a Kalashnikov and the opportunity to use it!" He laughed, touched by the idea of such devotion.

"Are there other Frenchwomen there?"

"Tons! There are a lot of French and Belgian women here. And I swear, they're almost worse than us. These days, they're obsessed with explosive belts."

"To scare people?"

"Yeah, and to blow themselves up if need be . . ."

I blinked.

"One last thing before I forget, baby. It's really important. Really, really important! When you get here, you have to cover yourself from head to toe—even your hands. The *sitar*[*] is mandatory here. You have one, right?"

** Slang for the additional veil ensuring that no part of a woman can be seen—not even her eyes.*

Thursday, 10 p.m.

Stunned, I looked at André. I'd managed to fake it in the past whenever Bilel mentioned unfamiliar aspects of Islam or Arabic vocabulary, but this time I was completely at a loss. What was a *sitar*? Kneeling on the floor in front of me with his camera, André gestured that he didn't know what it was, either. We hastily did a Web search on our respective phones—a dangerous maneuver on my part, since I was sitting in front of my interlocutor—and it came up empty. I figured it must be a slang word, guessing that for a fanatic like Bilel it might be the second veil a woman was supposed to wear, as dictated by certain radical laws. Holding my breath, I said I had one.

"Okay, that's good. If you respect the rules, you can go around by yourself with your girlfriends. But only when I'm not around. I'll take good care of you, I promise, but you should know that I have a lot of work, and sometimes I have to leave for several days at a time. You'll take good care of yourself while your husband is away."

"What do you mean?"

"You know . . . girl things . . . tricks to make your skin soft, for example."

Sometimes Mélodie's heart raced for Bilel. At other times he made her heart heavy with guilt for her "spoiled" life in the West. Paradoxically, the life he was trying to sell her was one of a princess in Syria, punctuated with visits to orphanages and hospitals. She felt oppressed by his demands. But Bilel had gained her trust, and she was determined not to contradict the one person who finally believed in her. Meanwhile, André and I were horrified by the hateful way Bilel was attempting to poison Mélodie's unconscious mind. And it had only been one week. We were disgusted by Bilel's libidinous innuendo directed at a woman half his age. All he wanted was to use her like a toy.

"I can tell you're a serious man, Bilel," Mélodie said, ignoring his insinuations.

"I've been doing jihad for a long time . . . but we'll talk more about that when you get here."

He talked about jihad the way a nine-to-five employee might talk about the company he works for.

"How long have you been in al-Sham?"

"One year. Before that, I did other things. . . . But I don't want to talk about it over the Internet; there are spies everywhere."

He winked at Mélodie.

"Were you in Libya?"

"Exactly! You're full of surprises, aren't you? You're really interesting, my baby . . . *bismillah*[*]."

---

* *Translator's note:* Bismillah *refers to an Arabic phrase,* Bismillah ir-Rahman ir-Rahim, *meaning "In the name of Allah."*

It wasn't surprising, though, considering most jihadists in Libya had fled, with their arsenal of weapons, to join ISIS's ranks. Bilel was being prudent, but I sensed he was itching to brag.

"I should keep some things for when you get here. I can't tell you everything now. You already know so much, and I sent you pictures of me in action. I can't wait to show you how amazing it is here! I'm so excited!"

We spoke a little longer, during which time I was able to glean a few meager details about him. I was beginning to put together the pieces of who he was before becoming the vengeful, power-hungry man he was today.

In another life, Bilel was called Rachid. Generally, he spoke very little of his early years. He was born near Porte de Clignancourt station, in northern Paris. His attendance at school was spotty, and he quickly dropped out. He claimed not to have kept any friends from the first twenty years of his life. A look of embarrassment crossed his face, and I couldn't tell if he was lying about his past or if his solitude had been such that he'd retained very few memories from that time. In truth, I had the impression that he didn't have many ties to his past. He claimed never to have married, insisting on his "fierce devotion to Allah." I suspected he'd tried and failed at various get-rich-quick schemes before finding his path to religion.

The police later filled me in on his extensive record. He'd started out with minor crimes, from dealing small arms to various types of theft. It was a way to get his hands on fast cash and acquire some local notoriety. He'd grown up Muslim, with Algerian roots, and become radicalized in the early 2000s. The security branch of the police force had kept its eye on him because he often traveled to Pakistan, a country overrun by al-Qaeda. But Rachid went there for religion classes on *tawhid*, a fundamentalist and dogmatic approach to Islam. As one agent told me when my investigation was wrapping up, "We can't watch everyone who spends time in these religious countries. They aren't all necessarily terrorists, and that's what makes this job so hard. Either people tell us we're incompetent, or they accuse us of Islamophobia." Domestic Security (the DGSI*) wasn't watching Rachid back then because it didn't have any concrete proof that he posed a threat to his country of origin.

Bilel confided his wish for children to Mélodie. Meanwhile, he had brought most of his family to the "holy land." Notably, his cousins. They were called the al-Firanzi, a name I wasn't able to verify, meaning "the French," and they were considered a powerful clan in Syria. However, he was very vague when asked questions about his siblings, and the way he talked about his immediate family struck me as disingenuous. Finally, I discovered in him a human and understandable feeling: loneliness. He was officially in charge of recruiting new devotees over the Internet. In reality, his proselytism was a way for him to combat his solitude.

It reminded me of the young jihadists who request interviews with journalists on Ask.fm. Editors often ask me to do such interviews. I've always turned them down. Such

** The DGSI (Direction Générale de la Sécurité Intérieure) is the intelligence service of the French police and is under the aegis of the Ministry of the Interior. The DGSE (Direction Générale des Services Extérieurs) is France's foreign intelligence agency, which depends on the Ministry of Defense and coordinates on missions abroad.*

conversations aren't worth much, since the jihadists tend to parrot nonsense they don't even understand. Now I realize that their motivation goes beyond fame. For them, such interviews also represent an attempt to attenuate loneliness. Like all teenagers, would-be jihadists communicate via text messages and express themselves using abbreviations. Unlimited text messaging was created for this age group. They have their own codes of communication and their own culture. They've grown up with new technologies. Bilel, on the other hand, was of an older generation known as the "big brothers." His generation had found what it was looking for—recognition and reassurance—in religion. When he talked about feeling completed by religion, his eyes glowed with a light I'd never seen in him before. Maybe I'm naïve, but at that moment he seemed sincere.

The impression didn't last long. Bilel rarely looked sincere. It's said that the eyes are windows to the soul. His eyes never reflected the peace I'd seen in those of true believers. The glimpses I got of his soul revealed his deepest desire: vengeance. It was up to me to discover why he sought revenge. What was it that made him "happy to have killed more than fifty thousand people"?

We hung up. André looked at me. He asked me to remove Mélodie's veil—at least the *hijab*. That night he found the costume especially disturbing. The night before I'd also rushed to take off my veil whenever Skype timed out. André gave me an affectionate smile, his expression marred by his rage for Bilel. He cursed him, and wished he could "punch that asshole in the face." André spoke of his twenties, when he lived in a bad Parisian neighborhood. Before becoming a photographer, he'd dabbled in delinquency. He'd even written a fascinating book filled with both serious and hilarious scenes about his adventures. Like Bilel, most of the characters in his book had run-ins with the police, mostly for burglaries and drug dealing. With the exception of those who'd died—accidentally or otherwise—André had kept in touch with everyone from his past. Some had converted and changed their lives. But no one had turned to radicalism, and certainly no one had left for *hijrah*. The thought of it drove André crazy.

"When we played at being big shots, it wasn't for a cause, and definitely not for religious reasons. And we never killed anybody! Abu Bilel is barbaric. He preys on the young and defiles Islam. He's a fanatic—a criminal! I mean . . . what an asshole!"

I didn't have to share André's life experiences to agree. As I took off Mélodie's disguise, he reiterated the need to wrap up our investigation as soon as possible. We would have a last conversation, during which I would fire even more questions at Bilel. Then I would erase all of Mélodie's Internet accounts, and we would publish the article. End of story. He was right, of course. But I still needed a few more days. My cover prevented me from asking direct questions. Mélodie was fragile, and Bilel strong. I wasted a lot of time playing along with Bilel's game of seduction in order to gain his trust. Embarking on this journalistic experiment had been a risk, and I would be frustrated if I didn't see where it led.

With the exception of a few editors, as well as Lou and Hadrien (both colleagues and friends), no one at the magazine where I'd pitched the article took my story seriously. I didn't take it completely seriously, either. The idea of me wearing a djellaba, using slang, and pronouncing a few Arabic words mostly made people laugh. No one, not even André, could comprehend the level of controlled schizophrenia that this exercise demanded. Sometimes I was so shocked by Bilel's words that I had to disconnect from Skype, but as I grew more accustomed to these exchanges, that became less and less necessary. André spoke to me—Anna—whenever communication with Bilel was cut off. I didn't have a second to change characters. My natural way of speaking came back, along with the reflex of grabbing a cigarette. I nervously rubbed my finger, searching for my favorite ring, which was never there, since I took care to remove it before each conversation. Later, when André saw the final report, his first reaction was "You really became somebody else. . . . I didn't realize back then just how dangerous this was."

A few days later

I spend most of my days at the offices of the two main periodicals for which I do freelance work. Back then, I could mostly be found at the magazine supervising my story. The staff is really tight-knit. Sometimes we get into rows, but essentially we're a big family. A tribe of tormented, highly dedicated, and passionate professionals. From young experts on the Belgian singer-songwriter Stromae to old-timers humming Charles Aznavour's "La Bohème," we really care for one another. Often blunt, we aren't afraid of words, and our skins get tougher as the years go by. We immerse ourselves in what are at times difficult and painful stories, and it brings us even closer together.

Over the years, everybody has worked with everybody, and we all know bits and pieces of one another's lives. The life of a reporter is different from that of a journalist. Reporters go into the field, following their instincts and conducting their own investigations, which is not something that journalists necessarily do. It's a difficult job, and the people who do it have to strike a balance between their emotions and professional distance. Sometimes we don't get affected. At other times, the story takes a little piece of us. We often travel alone, and while that isn't necessarily disagreeable, it's still strange, night after night, to find ourselves eating a plate of pasta alone in front of the TV, our only source of conversation the hotel receptionist. Whenever I knock on the door of a man whose wife has disappeared while jogging or of a mother who hasn't had any news from her son in weeks, I don't start by asking how they feel. I leave my pen and paper in my bag. Instead, I'm there. I think that's important. When people open their doors to me and let me enter their homes at a moment of crisis, we spend hours talking. And our conversations often have nothing to do with the reason I'm there to interview them. In my own way, I try to share their burden, and like others who do this job, I become a sponge for their grief.

Occasionally, the article doesn't get published, and I feel disappointed—not for my own wounded ego, but because I won't be able to share a story I thought was important. Luckily, some of us reporters still have the luxury of telling stories that we have really seen and investigated, without having to rely solely on the AFP.* Our editors know whom to assign depending on the subject matter. They know our strengths and weaknesses. For our part, we know whom we can count on. In my work family, I have the "young people's brigade," as our older colleagues have dubbed us. We're a well-integrated team. We offer advice and look out for one another. In this business, it's rare to have allies.

** Translator's note: Agence France-Presse, or AFP, is an international news agency based in Paris.*

I'm lucky to have precious friends, like Hadrien and Lou, whom I adore. They got sucked into young Mélodie's problems. As the days went by, they became increasingly concerned by the portrait I painted of Bilel. But there was so much going on in all of our lives that our laughter usually ended up masking our worries.

At lunch that day, I made a group of friends and colleagues laugh by telling them of the various habits and stories I'd developed in my role as Mélodie. I showed them pictures of Bilel acting macho and especially pictures of me as Mélodie. Their sniggering echoed through the lunchroom.

"You're so sexy in that veil!" Lou mocked.

"Does your fighter always wear so much eye makeup before going into battle? When he runs out, you can lend him your eyeliner!" Hadrien chimed in.

It felt good to be able to make light of things. Hadrien asked if Bilel really believed I was twenty. He's known me for a long time, and it was hard for him to believe that anyone would accept that I was so young. I answered that, strangely, the terrorist had never questioned my age. He didn't seem to care about how old I was. In fact, he acted so sure of himself, I don't think it even occurred to him to doubt my age. My friends asked more questions about the ruses I employed to be convincing. I told a few funny stories about André and his acrobatic feats as he tried to take pictures, without being seen, of Mélodie and Bilel chatting. But I avoided the romantic aspects of the conversation with Bilel and his attempts to turn this into a love story.

"How do you manage to video-call on Skype without being seen?"

"Actually, he can see me."

For proof, I took out my *hijab*, which I sometimes carried with me just in case. They laughed even harder.

"You're insane!" they cried, doubling over with laughter.

"What do you do when he says something in Arabic?" asked one friend.

I withdrew a giant yellow and black book from my bag: *Arabic for Dummies*, the ultimate weapon. That made them all choke with laughter. For the most part, Bilel and Mélodie spoke in French, but sometimes they used Arabic expressions. My friends made fun of my terrible accent when they heard me pronounce some of the phrases I used with Bilel.

We joked, even calling Bilel "my future husband." Laughing helped defuse the situation. As we headed back to work, Lou pulled me aside and told me to be careful. My stories were funny, but she had a bad feeling about this investigation. Before going back to his office, Hadrien gave me almost the same speech. He simply added that this had the potential to be a big story—if Bilel really was an important person. Hadrien didn't know just how high Bilel was in the Islamic State's hierarchy. At the time, I didn't even know, although I had been able to establish that he played an important role. Back in the office, displayed on my computer screen was the home page of an extremist site, *Shaam News*. The pro-ISIS source was full of useful, if slanted, information.

In the afternoon

After several hours spent checking into some of Bilel's claims, I was surprised to receive an email from Guitone, "the publicist." The message was sent to my professional account. Guitone knew he was speaking to Anna the journalist. He wasn't aware of Mélodie's existence.

Guitone was looking for news. He claimed to be "keeping an eye on intruders" from a café terrace and seemed to be bored. I imagined him perched atop a watchtower with a horn to blow in case of danger. It reminded me of the popular television series *Game of Thrones*. The two worlds share some similarities. As in the fictional story, the Islamic State defends its territories and tries to conquer new ones. Its fighters rape and pillage. For them life isn't sacred. They use spilt blood to legitimate their cause, but in reality what they are fighting for is land.

It occurred to me that since Guitone was French, he must know Bilel, in a kind of who's who of Islamists occupying Syria. Besides, he knew everyone. I pretended to have read an article on Abu Bilel and asked if he knew who that was. Guitone said he did. I rejoiced. Notably, he said: "Bilel is no joke. I have a lot of respect for him. He's taught us the guerrilla techniques he learned from the Chechens. He's an emir. He's also the French fighter closest to Abu Bakr al-Baghdadi." Guitone closed his sentence with a dozen exclamation points, underscoring the tight link between Bilel and Baghdadi. He couldn't have given me a better gift. I innocently asked if Baghdadi was ISIS's leader. I knew the answer, but I wanted to see how Guitone would present him. Yes, he said, Baghdadi was the head of the Islamic State. Guitone asserted that even he didn't know where Baghdadi was. But the leader watched over all. In fact, Guitone said, "he'll soon be the supreme caliph, as ordained by history." After that, he avoided my questions, preferring to talk about his new Nikes, since "they're really cheap in Syria." He asked me to mention that in my articles. I said goodbye to my correspondent and searched the Web for new information on Baghdadi.

I didn't find anything other than what I already knew: Abu Bakr al-Baghdadi was an Iraqi known under a number of identities, but his real name was Ibrahim Awad Ibrahim Ali al-Badri and he was forty-two years old. The American government promised ten million dollars to anyone with information leading to his whereabouts. I did learn one thing: *Time* magazine had just named him the most dangerous man in the world . . . and I'd just received confirmation that this paranoid and militant warmonger confided in a Frenchman, Bilel, who wanted to make me his bride. The day before, Bilel had told Mélodie that the two men had met in a city near the border of Iraq and Syria. I hadn't believed him for a second. Bilel, the French right-hand man of the Islamic State's leader—really? I thought of what Hadrien had said about what a big story it would be if it turned out Bilel was an important person in the Islamic State's hierarchy.

I took a deep breath. Everything was going to be all right.

I looked up from my computer, wanting to share this new information with any nearby colleagues. They teased me, and we shared a laugh. Meanwhile, Facebook kept sending Mélodie emails, notifying her that she had new messages from Bilel. The same two lines appeared a dozen times:

"You there?"

"My baby!!! Hello!!!!! Hello!!!!!"

I would reply later that night. Not in front of so many people. Telling them about our exchanges was one thing. Becoming the object of attention was quite another. I felt a little overwhelmed but not too worried. On the contrary. I'd worked hard to set this up, and I hoped Mélodie and Bilel's relationship would lead to some important discoveries. I couldn't give up now, since if I did, all the risks I'd taken would be for naught. I work in a microcosm, with reporters who've covered both the Persian Gulf War and the Arab Spring. I've seen them leave for war-torn countries, like others leave to take the metro, but equipped with bulletproof vests. I myself have covered frightening situations, on a more local scale: violent riots in France pitting the extreme left against the extreme right; anti-immigration—anti-everything—protests in France, Turkey, and elsewhere. Still, the risk I ran here in Paris seemed pretty small compared to what some of my colleagues faced. I knew this investigation was dangerous, but I didn't really feel threatened. I think it's because there was still distance between Bilel and me.

My various conversations that day reminded me of a famous quote by Michel Audiard* that my older brother used to like reciting: "Two intellectuals sitting down will never go as far as one simpleton walking."

* *Translator's note: Michel Audiard was a French screenwriter, film director, and author whose career spanned from 1949 to 1985.*

The same day, 5:30 p.m.

As we left the paper at the end of the workday, André again reminded me to wrap up this story as quickly as possible. He'd been checking in on me, concerned, since that morning, and was bewildered by my calm. I was touched, and I knew André was right. If Baghdadi was really as close to Bilel as I suspected, the sooner I could finish my investigation the better. I'd informed the editor supervising the project that Bilel might be linked to Baghdadi, but I'd refrained from mentioning anything to André. I was overcome by the reporter's thirst for knowledge; I had to see this story through to the end, but I was beginning to sense that this impulse was weakening my instinct of self-preservation. Bilel was a mine of information. I had already learned a lot about ISIS and the mentality of its members. I told myself it was worth continuing the investigation, just a little longer.

André advised me to ask more direct questions and to keep our conversations short. I argued that might awaken Bilel's suspicions, and if I wanted David to slay Goliath, then Mélodie would have to lend the terrorist a patient and attentive ear. It would take more than one last conversation.

Although Bilel was probably manipulating Mélodie's emotions and using her for his own ends (as he undoubtedly did with all his prey), he was fond of Mélodie. Over the past few days, I'd sensed his growing impatience to see and speak with his betrothed. It wasn't a game for him. If Mélodie ended up in Syria, she and Bilel really would marry. He wouldn't make her into a sexual slave. Not that I think he would be a good husband to her; I'm not that naïve. Despite his claims to the contrary, I knew he probably had at least one other wife. Still, he viewed his nighttime conversations with the young Mélodie as a reward at the end of a day in the line of fire. He fell asleep thinking of her. I don't dare imagine where his thoughts wandered. But his feelings for Mélodie, together with his conviction that he was manipulating her, gave me an advantage. Bilel was distracted, and Mélodie was the one really running the show; she saw right through his tricks and turned them to her own advantage. I still needed my digital avatar. A hasty attack would endanger my investigation. I had a vacation planned with Lou at the end of the week. I'd wrap up the investigation then.

At seven o'clock sharp in Syria, six o'clock in France, Bilel was waiting for Mélodie at his screen. He was an hour early and seemed particularly well groomed. This militant fighter looked more like a pretty boy obsessed with his own "hotness" than a soldier. The gap between Bilel's fanaticism and his teenage mannerisms was jarring, even if nothing about him or his fellow fighters surprised me anymore. At this stage of my investigation, I was beyond laughter or even tears. Bilel sat slumped in a tattered chair, the Internet café behind him depressing and empty. His eyes brightened when he saw his intended, and he instantly straightened, adopting his usual pose of the swaggering man of power. He tilted his head back as he put on his gold-frame mirrored Ray-Bans, which covered most of his face. At this hour, Syria was already bathed in darkness and the room around him was somber. He wore a dated bomber jacket. His general look reminded me of Starsky, from the seventies television series *Starsky & Hutch*. Bilel told Mélodie about his day. In return, she fretted about the dangers of his "work." She was scared for him. André rolled his eyes. Bilel reassured Mélodie: he was brave, experienced. *Nothing* scared him. Besides, he wouldn't have gotten where he was in the organization if he hadn't proved himself to be a good strategist. I sensed he'd wanted to say "excellent strategist," but his false modesty had stopped him. Bilel was so humble; Mélodie couldn't help admiring him. She wanted to know more about him. With a little prompting, Bilel described a few new aspects of his daily life.

Depending on the length of a battle, often fighters can only squeeze in a few hours of rest in their cars. They sleep between two and five hours per night. They spend the rest of their time between Raqqa, which is ISIS headquarters and a bad place to live unless you've pledged allegiance to Baghdadi, and a neighboring town about a dozen miles away. No, Bilel wasn't in Aleppo anymore, but for Mélodie's safety, it was better if she didn't know his exact location.

ISIS strictly enforces sharia law in Raqqa. Bilel liked the firm observance of his religious dogma. Life in Raqqa, according to Bilel, was beautiful and free, with its cafés, movie theaters, and shops. He and his men had liberated the city;

its citizens were grateful and always showed them respect (in reality, almost three-quarters of its residents want to flee the city, but ISIS and its laws have barred them from leaving). Bilel was an officer, and the Islamic State was his army. Unconcerned by the contradiction, he explained that women in Raqqa had to wear a full-body veil and were not allowed to go out, except at designated times and in the company of their husbands, fathers, or brothers. Those were the only constraints, and according to him, they weren't in the least constraining. In the streets of Raqqa, ISIS fines husbands accompanying improperly veiled wives between $85 and $225. If he doesn't pay for the "infraction" immediately, his wife will be killed. In contrast, the fine for men not wearing a beard or djellaba is approximately $35.

Residents of cities seized by the Islamic State have no way out. Their only option is to "get in line or die," the grim refrain of dictatorial regimes. ISIS uses one of the five pillars of Islam*—*zakat*—to justify its demands for money. *Zakat* is usually defined as a tax, without much further explanation. In reality, it symbolizes the fact that Muslims are expected to help those in need. As another god says in the Old Testament: "Love thy neighbor as thyself." *Zakat* therefore has nothing to do with the length of a beard or a sloppy use of the veil, and it has even less to do with an organization that amasses millions every day from oil and seizes money from the destitute in order to fund terror.

* *Declaring one's adoration for one God, and recognizing Muhammad as God's messenger; ritual prayer five times a day; fasting during the month of Ramadan; paying an annual tax for the poor (zakat); making a pilgrimage or religious journey known as the hajj to Mecca, in Saudi Arabia, at least once in one's lifetime.*

Bilel was not allowed to keep his cell phone in Raqqa. It could be traced. Besides, the network was patchy. His only means of communication was through a walkie-talkie. He also had very spotty access to the Internet, except in cafés, where anyone could hear our conversation. Bilel therefore got up every morning at six o'clock and found a quiet place from which to send Mélodie a tender message. "Have a good day, baby. Think of me. I miss you." Followed by a bunch of red hearts. The first few times I received such messages, my blood ran cold. Then I learned to laugh them off with André and some of my trusted colleagues. But after a while I found his smothering behavior grating. At times I wanted to smash my computer against the ground. Instead, I rolled my eyes.

Bilel drove his pickup, filled with guns and bottles of chocolate milk, which he was crazy about, for hundreds of miles each morning "for work operations." He didn't provide much more detail. Although he probably sometimes went to sensitive areas, I was now sure he spent the majority of his time in Deir ez-Zor, near the Iraq border, about two hundred miles from his headquarters. He told Mélodie he often stopped to give orders to his French battalions. He made decisions about their daily priorities. He said who would go to the front, who would police the cities controlled by the Islamic State, and who would take care of the infidels "whose time had come." It was also up to him to organize religion and language classes, since there were many new arrivals who didn't speak Arabic. In addition, the mujahideen in Syria spoke different Arabic dialects. Each nationality had its own form of Arabic, making communication difficult.

Bilel was an emir, and he had to anticipate situations and problems. To Mélodie, he confided what was exchanged during "secret meetings" with Iraqis or members of al-Qaeda seeking to switch sides. Such meetings mostly took place in hidden tunnels linking Syria and Iraq. Bilel claimed that he had a map of these tunnels and knew every nook and cranny by heart. He had even dug some of the tunnels himself. Bilel's mission during these meetings was to "negotiate peace," because he was "the highest-ranked Frenchman and the man closest to al-Baghdadi." For the first time, he began to speak to Mélodie about his ties to the "caliph," confirming what Guitone had told me.

Al-Baghdadi had recently sent Bilel to meet with Abu Mohammed al-Joulani. Bilel mentioned that tidbit innocently, bragging, to Mélodie, who didn't understand. But for me this was precious information. Al-Joulani is the head of the main Syrian brigade of al-Qaeda. Amid the clashing militant organizations in Syria and Iraq, he has emerged as one of the main figures of Middle Eastern terrorism. If ISIS had sent Bilel to "negotiate" with al-Joulani, an almost invisible man, it meant he must have an important role in the organization. Mélodie casually asked how the meeting had gone. Bilel boasted that the two men had reached an agreement. Soon, he said, they would declare a caliphate. But who would be in charge, the Islamic State or al-Nusra, the main Syrian branch of al-Qaeda? Mélodie asked, showing her erudite interlocutor she'd remembered his teachings. Bilel's face tensed. It was an important question, since each clan hoped to be the one to inscribe its name in history. He explained that they had agreed on the main point, that Syria would become an Islamic state. Al-Qaeda, which mainly focused on building cells to target the West, would have to ally with ISIS in Syria. He was avoiding the question.

Mélodie quickly questioned him on the issue of black gold. As usual, the fanatic dodged the subject of ISIS's lucrative oil refineries. He was liberating a people; he had to defeat evil. Mélodie admired his courage. Although she didn't completely grasp his utopian aims, she reasoned that his ideology was fair and noble. This man knew much more about life than she did. She was impressed. She asked how many fighters were in the Islamic State.

"We don't call it the Islamic State here, but al-Dawla al-Islamiya fi al-Iraq wa al-Sham! There are at least ten thousand of us in the army."

"Really? That many?"

"Even a little more. At the rate things are going, six months from now there will be fifty thousand of us."

"At the rate things are going?"

"New fighters arrive here every day. A lot of French people, Belgians, Germans. A ton of Tunisians. Not to mention the local Sunnis who've joined our cause, and brigades from all over the world, like Boko Haram,* which has allied with us."

His projected figures were slightly exaggerated, but the rest was tragically true.

Bilel was in a sharing mood that night, and Mélodie took advantage.

"When I don't have any news from you, I try to learn about where you are and what you're doing. I heard somewhere that you guys are really well organized. How does it work?"

** A Sunni terrorist organization based in Nigeria. Its name means "Western education is forbidden." Its leader, Abubakar Shekau, was behind the kidnapping of 273 young girls on April 15, 2014, which led to a global campaign, "Bring Back Our Girls."*

"Everyone has a job. When you get here, unless you already have experience, you follow a basic curriculum: language classes in the morning and shooting classes in the afternoon. You sleep in a *katiba*[*] with other Francophones, mostly, and experienced fighters committed to guiding you in your spiritual journey. After two weeks, either you're strong enough to fight and you join combat or secret operations, or you specialize in a specific field, like recruitment or counterespionage. You can also devote yourself to noble activities like visiting wounded jihadists in hospitals or delivering medicine to those in need. You can teach the Koran to the unenlightened. The rest of the time, you're free to do what you want. Life is beautiful here, and it's cheap. We're fighting for freedom."

Freedom. Bilel's idyllic description of life under ISIS neglected to mention that most jihadists were subalterns.

"But I don't have money," Mélodie replied pragmatically. "Life may be cheap, but according to sharia law, I don't have the right to work. How am I supposed to get by?"

"It's different for you. You're a woman, and my future wife. *Insha'Allah*. The organization gives all believers a monthly salary of fifty to two hundred and fifty dollars.[†] You'll have more money in Syria than in France! You'll be rich. In France, you're fucked; here, we fuck up French nonbelievers. Anyway, men and women don't have the same jobs. I was just telling you about the men's training program."

"I heard that the people who become suicide bombers

** Residences or buildings requisitioned by the Islamic State.*

† The average monthly income in Syria is $218.

only do it because they want to come back to Europe and know there's no way out of Syria—at least not without going to prison. Is that true?"

I was thinking of Nicolas Bons, a young man from Toulouse who converted from Catholicism to Islam. He became famous when he and his younger brother made a video asking French president François Hollande to convert to Islam. Smiling, he invited Europeans to join him in Syria to do their jihad. He'd been the one to recruit his brother, who died a few months later, as cannon fodder at the age of twenty, for a cause that wasn't his own. Soon after, Nicolas, who had been working as a language teacher and European recruiter, volunteered for a suicide mission. He blew himself up in a van near Aleppo.

Back then, I found the last photograph of him, taken a few minutes before his death. In the picture, he's pointing his index finger to the sky, a common gesture among Muslims to indicate God. The smile on his face looks nothing like the one he wore in his video to the president. His eyes are empty, or more precisely, they're filled with disillusion.

After their deaths, I met the boys' father, Gérard Bons, the head of a successful solar panel company. He'd been living in Guiana for his business for the past several years. He agreed to meet with me on a relatively rainy day in the lobby of a Cayenne hotel. He shook my hand firmly, and as I studied his face, which seemed permanently frozen in grief, he said: "I warn you, I'm only granting you twenty minutes. I'm not interested in voyeurism, tears, or sensationalist journalism." Less than five minutes later, when I brought up his only living son's remorse over his brothers, he broke into tears. I'd met the youngest son the day before, with his group of friends, the same as his brothers'. He'd been in a state of shock and was filled with guilt. I've personally lost many people dear to me, and I know how important it is to manage feelings of guilt before they harden and become impossible to dissolve. Instead of speaking to this father of his dead sons, I tried to convince him to focus on the living. Namely, on this little brother. I shared some of my own painful experiences, offering advice based on what had helped me. I knew it wouldn't necessarily be of much help to him. Our lives weren't the same. But it was the least I owed this man, this father who had agreed to talk to me about losing not one but two children. And this modest man gave me his clear and poignant testimony. He'd begged Nicolas to return, but the answer had always been the same: "I want to . . . but it isn't so easy to leave Syria, Dad . . . and even if I managed it, once back in France I'd be thrown in prison." Gérard Bons told me that Nicolas felt responsible for his younger brother's death. According to him, he'd deliberately killed himself in order to put an end to his overwhelming sense of guilt.

I barely had enough time to think of that honorable but broken family, before Bilel quickly replied to Mélodie: "Of course that's not true. Suicide bombers are our strongest fighters. We evaluate strength based on two things: faith and courage. A person brave enough to blow himself up for Allah will go to paradise, with honors."

Some suicide bombers might indeed be belligerent enough to sacrifice their lives, but generally, at least in ISIS, the weakest members take care of material support (they chauffeur or prepare meals) and the "slightly less weak" blow themselves up. One more, one less. What difference does it make to them? Their ranks are growing every day.

"You tell me every day that you only care about one thing: paradise. Why not become a suicide bomber?"

He took a moment to reply.

"I'm still needed here. It's not my time yet, *insha'Allah.*"

"You've been telling me about new arrivals, but how does it work for people like you with more experience, older people? You're thirty-eight, right? On the news, they always talk about minors and young adults leaving to fight."

Bilel was annoyed.

"How do you know I'm thirty-eight?" he asked sharply.

I wanted to tell this idiot, who claimed to be a genius in counterespionage, that all his information was listed on his Skype profile. City: Raqqa. Nationality: French. Age: thirty-eight. On a personal level, I could never forget the year he was born, 1976, the same year as one of my brothers. One of life's ironies: I have that date tattooed in minuscule on the inside of my right ring finger. Luckily, they weren't born on the same day. Mélodie replied that she'd seen it on the Internet.

"Yeah, but I don't look my age," he bragged. "Everybody thinks I'm twenty or twenty-five. I have good genes. Besides, age is just a number. You should see all the European women who want to come here for me. I'm really attractive, baby."

I couldn't believe him! Bilel was now acting as if he were a rock star. André rolled his eyes. I sensed he was struggling to contain his irritation. Internally, I mused that perhaps Mélodie should send her sweetheart some antiwrinkle cream to maintain this "metrosexual jihadist" looks. I would have toyed longer with this new aspect of the modern jihadist—it was tragically fascinating from an anthropological point of view—but I sensed André wanted me to wrap up our conversation.

"What's the training program for girls, then?" Mélodie asked, changing the subject. "Is it different for converts?"

"We jihadists prefer converts," he said, laughing.

André and I exchanged a brief look of surprise.

"Why?" Mélodie asked.

"*Masha'Allah!* Because you're more serious about religion, and at the same time more open about life. You're not like these Syrian women who wear the veil but don't know how to make their men happy. *Insha'Allah.*"

Bilel had just betrayed himself. He who was always bragging about the Syrian people—and congratulating himself for freeing them—had just insulted them.

"What do you mean, we're 'more open'?"

"You know what I mean. . . ."

"No, I don't."

"You're more . . . affectionate, if you know what I mean."

"Still no."

"You're more imaginative . . . with your husband."

"Isn't it *haram* to be 'more imaginative' in that domain?"

"You can do what you want with your husband when you're alone with him. You owe him anything and everything. But only him. You need to fulfill his every wish. You can wear whatever you want underneath your *sitar* and burqa. Garter belts, fishnet stockings, anything your husband might like. Do you like pretty lingerie, baby?"

Monday, 7:30 p.m.

I disconnected. I couldn't think of an answer, not on the spot like that. Mélodie was an improvised creation. I was making her up as I went along. I had given her a backstory that even included some romantic details, but I hadn't considered her erotic self. I felt like I was suffocating inside my thick black veil. Frustrated, I took it off, drank a large glass of water, and lit a cigarette. In that moment, the human being in me overtook the journalist. The terrorist had trapped me, and I was furious with myself for being so preoccupied with other issues that I hadn't considered how I might respond if he led the conversation in that direction. I looked at André, who was cursing and pacing through the living room like a lion in a cage.

"Who does that insane pervert think he is, to ask you what kind of lingerie you wear? First, he practically orders you to go to Syria, then he wants you to marry him, and now this talk about your garters! What next? Will he ask you to strip for him? For the sake of Allah? I hate this guy."

"Me, too, but let's calm down." I needed to think of something to say, fast. Bilel was calling, and if Mélodie didn't answer, he could grow suspicious.

Pretending to adopt his medieval ideology, Mélodie whispered faintly, as if corseted, "I'll wear whatever pleases my husband, but since I'm not married, I can't discuss this with a man."

"That's good. I knew you were pure, Mélodie. Before I saw your face, I knew you were beautiful."

"But you were the one who taught me that religion isn't concerned with beauty."

"That's true. But you and I will have extremely beautiful children . . . *insha'Allah*. You look curvy, which I like. And, as I said before, I'm good-looking."

Bilel was again ignoring Mélodie's words. He bit his lip and silently stared at her. I lowered my eyes and waited for the awful moment to pass. From where he sat, André couldn't see Bilel, who was slowly licking his lips. How was I going to get him to talk about Baghdadi—or anything else—now? I gritted my teeth. I would have to be patient.

"Am I your type?" he asked.

"I don't really have a type."

"But you said you thought I was hot! So if you don't have a type, allow me to ask you again to be my wife."

"But I don't know you very well, Bilel, and I'm scared. If I say yes, that means I promise myself to you for the rest of my life. But what if you have other wives?"

"Listen to me, you are my jewel, and the house where we'll live with our children will be our kingdom," he said, ignoring my question. "Come live with me, and you'll see you can trust me. Can I ask you a question?"

At this point in the conversation, I was ready for anything.

He tended to jump from one topic to another, with no apparent logic.

"Do you have long hair?"

"Yeah . . . why do you ask?"

"Really long? Or just mid-length? Because most girls say they have long hair, but they're lying; it's really just mid-length."

I had no idea where he was going with this. Glancing at André, who looked equally baffled, I had to keep myself from laughing. Bilel easily shifted from boasting about his gruesome "exploits" to deploying his pathetic techniques of seduction. It felt like being on a dating website, "adoptajihadist.com."

In general, I try to look for the good in all human beings. I like to think that there are solutions to the complicated situations life throws at us. When it came to Bilel and this situation, I was at a loss. But his obsession with the length of Mélodie's hair was an excellent antidote to the more disturbing aspects of our conversation.

"My hair touches the middle of my back."

"So it's mid-length—not long."

"So what?"

"Nothing, it's just, I love long hair. Is it curly?"

"Not really, just wavy."

"That's perfect. I asked Allah to put a converted, green-eyed brunette on my path, and here you are, Mélodie. My wife—"

"I haven't said yes. . . . I should go. My sister just got home, and I have to take off my veil or she'll tell my mom."

"Take your computer into your room, my wife. I'll wait. But swear on almighty Allah that you love me."

"But I share a room with my sister. I really have to hang up, Bilel."

"Okay, but I want you to know that I'll go to sleep tonight with the certainty that Allah sent you to me."

"Okay, Bilel. *Masha'Allah*. Sleep well."

"Mélodie . . . don't forget, you're mine now. You belong to me for eternity. Do you understand? Don't forget!"

Monday, 8 p.m.

I closed my computer. André joined me on the sofa and solemnly handed me a cigarette, which I quietly smoked. For the first time, I wondered if I was becoming a little schizophrenic, torn between moments of comical absurdity and others of suffocating fear. After a while, André broke the silence.

"This has gone far enough. You're going to stop this now. Done! Over! *Khalas!** Do you realize how risky this is? Tomorrow will be your last conversation with Bilel, and then we'll make Mélodie disappear. Got it?"

I remained mute. I agreed with him, but I didn't know what to say. I felt like a tightrope walker teetering forward on an invisible wire.

Milan was due to arrive. At least I'd finished in time. I wouldn't have liked him to see me like that, in somebody else's skin. Milan wasn't familiar with what I was working on. All he knew was that I had regular cyberdates with a French jihadist in Syria. I felt exhausted and depleted of

---

* *"Enough!" in Arabic.*

energy. The intercom rang, announcing Milan's arrival. I asked André to avoid discussing our investigation in too much detail, and went to open the door. As I came back into the living room, I heard the Skype ringtone coming from my computer. Bilel was calling back. Answering was out of the question. But André, thinking Bilel might have something important to say, urged me to put Mélodie's veil back on. Why else would he be calling? He knew Mélodie couldn't talk right now.

I hesitated. I heard Milan take off his helmet in the stairwell and quickly climb upstairs. I felt suffocated. Without thinking, I hastily drew the veil and djellaba over my clothes. I clicked the green button just as I heard Milan enter and close the door to my apartment. In a few seconds, the time it took to walk down the hallway, he would see me. It would be a shock for him, and I hated that I couldn't even look at him. André signaled for him to keep quiet and stay in a corner of the room. I tried to concentrate on what Bilel was saying, but I was also looking at my real boyfriend from the corner of my eye. As soon as our eyes met, he lowered his gaze. He didn't recognize me. Or maybe he did and he couldn't stand the sight. He went to the window and lit a cigarette—he almost never smokes. I was embarrassed to have to flirt with another man in front of him. It didn't matter that I was doing it for my job. In some ways, that made it worse. Never in a million years would I have imagined myself doing this—and certainly never in front of the man I love. I had neglected to mention to him the dark details of what my avatar demanded of me; what else might I be hiding? Nothing, of course, but I was worried he might not trust me anymore. Or that he would fall asleep that night with all kinds of terrible questions in his head.

"I wanted to say good night again, baby."

Awesome. Bilel didn't have anything more to say, aside from another smothering expression of his growing "love" for Mélodie. I wished I hadn't listened to André. I was disgusted to have to do this in front of Milan.

"Okay. Good night, Bilel," Mélodie replied coldly. "I told you I can't talk right now. I listen to your instructions, please listen to mine. If my sister comes into my room, I'm dead. I'm hanging up."

"Okay. Sweet dreams, beautiful, and don't forget: you're mine forever."

He hung up. There had been many long silences in my living room since these conversations had first begun. André and I had heard Bilel say terrifying, unspeakable things before. But for the first time, his words really made me flinch, because, without realizing it, Bilel had found a weakness in me, the journalist. He was speaking to Mélodie, but his words attacked my values, my convictions, and my idea of humanity. And with Milan present, he had just inserted himself into my personal life. Worst of all, it was my fault. I was awash in emotions: embarrassment, regret, anger, and others I couldn't even name. Everything was happening so fast. Tearing off my *hijab* and djellaba, I morphed back into myself and rushed toward Milan at the window. I threw my arms around him, hugging him from behind and murmuring an apology. He remained motionless, nervously inhaling his Marlboro Light like someone unaccustomed to smoking.

As André packed his equipment and prepared to leave, he babbled uneasily and tried to downplay the significance of the conversation. After he left, Milan handed me a motorcycle helmet and asked me to sleep over at his place. I said, of course. I completely understood what he was feeling. The first time I'd spoken with Bilel I hadn't wanted to sleep at my place, either. Milan peered at my miniskirt and T-shirt of the Clash. He seemed to be looking at me differently than before. He said he was worried about me. He regretted not knowing more about what was going on in Syria, but it seemed obvious to him that I was putting myself in danger. He wished I'd talked to him more about this investigation. Now that he'd seen me as Mélodie, he didn't want any more details, that is, not unless I encountered any problems or had to travel. That suited me. The rest of the night was ours. Still, I felt guilty for having involved another person in this story. As they say, "Ignorance is bliss."

I fell asleep feeling embarrassed and contrite, as if I'd been caught having an affair. Milan put his arms around me, but it felt more mechanical than tender. That night, for the first time, Mélodie, my avatar, had affected my life. She may have been virtual, but her impact was real.

Two days later

Mélodie's list of virtual friends grew as soon as Abu Bilel announced his plan to marry her. Her recent posts on Facebook calling for "humanitarian" jihad elicited new "friend requests" and private messages. Girls, all called Umm something-or-other, who treated mujahideen like stars and specified on their profiles that they wouldn't accept male friends, began asking Mélodie for advice on the safest route to al-Sham. She received messages in French, Flemish, Arabic, and I think even German—a true melting pot of languages. My friends who speak Arabic fluently floundered over the variety of dialects used in these messages. Some of the questions asked in the messages were both technical and strange: "Should I bring a lot of sanitary pads or can I find them there?"; "If I arrive in Syria without a husband, it's probably not a good idea to draw attention to myself by bringing thong underwear; my future husband might think I'm immodest. But will I be able to find them there?" I was bewildered by the fixations of these girls who were signing up for death. How was I supposed to answer their questions?

I spent my evenings coaxing Bilel to talk. He was

becoming needier and wanted to see Mélodie all the time, so sometimes we spoke without the hidden presence of André. In my professional life, I work on a range of topics, not just those pertaining to ISIS and Syria. Now I was devoting all my time to this investigation, ignoring other current events, like what was going on in Ukraine. Before meeting up with Milan, or whenever he was busy, I put on Mélodie's *hijab* and tried to connect with her suitor over the Internet. Most of our nighttime conversations took place at my apartment, with me wearing my veil, cross-legged on the sofa. I didn't want to take advantage of the girls writing to Mélodie. Young people are often impetuous and fragile, and on moral grounds I decided not to reply to their messages. Besides, it would only lead to fruitless debates. However, I did respond in cases of imminent departure. I drafted a copy/paste message in which Mélodie discouraged girls from leaving and sent it to all applicable correspondents. Here it is without the grammatical errors and emoticons:

> *Salaam, sister,*
>
> *Like you, I too lost hope and felt helpless in my country, whose laws are often incompatible with our own.* Islam saved me from disillusion and despair. But being a good sister is first and foremost about behaving with dignity and discretion, and examining the lessons laid out in the Koran. It's not about watching videos advocating nonsense. To do jihad*

---

* *Islamic laws.*

is to work on the self and respect our religion's precepts. It is to treat your neighbor—Syrian or otherwise—with goodness. Leaving to prove something doesn't do anyone any good. Open your eyes: there are people all around you, your brothers and sisters, who need your help. If you feel lost, do as I have done, talk about it with your parents (imagine: I'm a convert), and think of the grief you will cause them if you leave. I once planned a trip like the one you're considering. I was in contact with a lot of jihadists. I was convinced I was doing the right thing. And then I followed the advice I've just given you, and today I'm happier than ever.

I know these girls were drowning, and that my message was only a drop in an ocean of desperation, but I couldn't do nothing. Perhaps without realizing it I was trying to clear my conscience.

French-speaking male jihadists between the ages of sixteen and twenty-seven also sent Mélodie gracious messages. Their first three questions were always the same.

"Are you in al-Sham?"

"How old are you?"

"Are you married?"

I didn't have time for these mostly fruitless exchanges. I did, however, keep in touch with Abu Mustapha, a twenty-seven-year-old Frenchman living in the Islamic State who seemed more intelligent than the others. More honest, too. Mustapha has been devoted to his religion since he was a young boy. He's familiar with the history of religions, from their origins to the present day. He lives his life for his god, because that's what makes him happy, and he doesn't try to convert the people around him. He doesn't call for bloodshed. He knows that his jihad is first and foremost a personal spiritual journey. He's like a Christian on a pilgrimage to the Vatican. Jihad doesn't have to imply war. *Hijrah* does. Al-Qaeda, ISIS, and other factions have sold their own versions of it.

In search of himself, Abu Mustapha went to the Middle East for religious reasons. He has not yet become a murderer. He never posts pictures of himself in the region, nor does he share propaganda slogans. Nothing of what I know of him would suggest he's a fighter. His Facebook updates are usually pretty verses from the Koran, and he only preaches one thing: to be true to one's faith if one is a believer. He has always been strictly devout. When he went to Syria, he expected to find a war-torn country that he and his new brothers would liberate and transform into an Islamic State. That's what he imagined when he left everything behind. His disillusion since his arrival has been profound. He confided to Mélodie that sometimes he felt like he was "living a lie." Although we only ever communicated in writing, I sensed he was deeply lonely. Mélodie suggested he return to Europe. Or maybe, if it was still possible, he could try to start a family far away from the violence.

"I've been religious since I was little. My family is very pious. They're not ashamed of me for going to Syria, because they know my heart is pure. But they're afraid of ISIS, al-Nusra, Bashar's army, and anyone who kills in the name of religion. They want me to come home . . . and even if a good Muslim shouldn't be afraid to embrace death, sometimes, sister, I fear my time is up."

"Is it difficult to adapt when you first arrive? Do you miss your family?"

"At first it was really hard. You miss your friends and especially your family, *insha'Allah.* My little sister graduated from high school shortly after I left. And most of my brothers have had birthdays since then. I missed out on all those moments. I've been crying for a year, sister."

I was sincere with Mélodie's reply:

"I'm really sorry. . . . I know it's very hard to leave Syria, and your troubles aren't over even if you do make it back to Europe. But nothing ventured, nothing gained. It will be difficult at first, but if you can prove that you never took up arms, there are a lot of NGOs that would be interested in helping you."

Abu Mustapha took a long time to reply. I didn't know him, and yet I waited on his words with bated breath. I was already dialing Dimitri Bontinck in my head and asking him to help extract him. But when he spoke, he repeated ISIS propaganda, like a broken record. Still, I'm convinced he really considered Mélodie's advice.

"Change can't happen without revolution. Suffering and the loss of human lives are necessary evils. I pledged allegiance to our future caliph, Abu Bakr al-Baghdadi. He and he alone can guide us Muslims. I made it here. I've lived here for a year. I can stay here for the rest of my life. *Bismillah.*"

"If I joined the Islamic State," Mélodie said, "I would only do humanitarian work and I'd start a family with a man I loved. I wouldn't get married simply to adhere to the prescribed model."

"Are you married? Do you have a fiancé?"

"I have somebody waiting for me there, yes."

He didn't reply for several minutes. He was probably disappointed.

"I'd like to start a family. I'd like to become a father and find a wife I love, but it's complicated here in Syria. It's difficult here, sister. The mentality of Syrian women is completely different from ours. That's why we prefer marrying European sisters."

"Why?"

"Because Syrian women look down on foreign jihadists. They're afraid of ISIS, and they don't think we're compatible with their faith. They don't follow sharia law like you European women do. They won't even wear the *sitar*, opting instead for tiny *hijabs*."

I thought of the first time I had to put on my *sitar*. It seemed so distant now, but it hadn't been that long ago. Abu Mustapha continued.

"Also, I grew up in France, and they grew up here. The cultural gap is too wide. They're closed-minded; they don't understand our Western ways. That's why life would be perfect if we could marry sisters like you."

"They're closed-minded?"

"Yeah, they aren't open to anything—not religion or their husbands."

"But I heard a lot of European women have been arriving in Syria."

"They don't grow on trees! You have to look hard."

"Here there are new stories almost every day about people leaving for the Middle East, and it's often women. I personally know a lot of sisters who have left."

"European women come here if they have a husband waiting for them or if they're truly interested in doing their *hijrah*. But fewer women than men actually make the trip. They don't have the courage. Most often when they do come, they're lured by materialism and think they'll be treated like princesses. When they get here, they're surprised and terrified by what life is really like, and they get depressed."

Before replying, I considered what Bilel had told Mélodie about life in the Islamic State.

"I've heard there are some cities, like Raqqa, where you have access to the same products and technologies as we have in the West, so long as you obey sharia law."

"Not really. Well, it depends on who you are. ISIS is really well organized, and the more important you are, the better your living conditions. But I can assure you that life here isn't like it is in Paris."

"Are you from Paris? I live in Toulouse."

"Yeah, and although I couldn't always express my religious beliefs, I don't blame France. All I wanted was to be able to live according to my beliefs and the laws of my religion."

Abu Mustapha wasn't Bilel. He was lost, too, but he hadn't lost sight of who he was. Not completely . . . He struggled to make sense of today's contradictory visions of Islam and of its religious practices. In the midst of that confusion, he'd joined ISIS, an organization whose only aim is to impose its tyrannical ideology by all necessary means. If

Abu Mustapha one day managed to return to France, his membership in that organization would be damning.

Jihadists returning to their home countries are first placed in police custody, before being put under house arrest or provisionally detained while the authorities determine whether or not they represent a threat. In all cases, the law is clear: they are charged with "conspiring with a terrorist organization." The issue of returnees is a thorny one, since it's extremely difficult to distinguish between a person who regrets having been influenced by the Islamic State and its propaganda, and a fanatic who might be planning an attack, as was the case with Mehdi Nemmouche, a former ISIS jailer who shot and killed four people at the Jewish Museum of Belgium in Brussels in May 2014, shortly after his return to Europe from the Middle East. This shooting was the first of what he had planned to be a series of similar attacks throughout Europe.

The number of departing minors has increased since January 1, 2013, with the implementation of a new law allowing French teenagers, fourteen and up, to travel throughout Europe without a signed authorization from their parents or legal guardians. You think your kids are at school, when actually they've boarded a one-way flight to Turkey. Authorities are trying their best to handle the situation, but they're overwhelmed by the alarming increase in the number of departures. Considering the risk of attacks in France, failure to stop these departures can have terrifying consequences. French minister of interior Bernard Cazeneuve is working to stanch the outflow of minors by identifying and stopping potential departees. He also takes the threat of those who return to France very seriously. But Cazeneuve's approach to this task is more police- than rescue-based. And what of all the jihadists bragging about traveling back and forth between their home countries and Syria? Mélodie was Facebook friends with a fighter who called himself "If you want my opinion"; he posted lots of pictures taken in Marseille, his birth city, where he went "to see friends," and of himself, proudly dressed as a fighter. He was unafraid and unapologetic, captioning a picture of himself holding a Kalashnikov with these words: "Your possessions are a reflection of who you are." Three days later, "If you want my opinion" shared pictures of himself in Syria, wearing brand-name European clothing from head to toe.

Bilel often hinted that returning to France was easy. But that was only the case if one's stay was short and the itinerary difficult for French authorities to trace.

That night, Bilel demanded to know Mélodie's "new Islamic name." In his mind, he and Mélodie would soon be married, and he wanted his future wife to choose the name she'd have in her new life. In the last few days when Bilel had asked this, Mélodie had avoided the subject. "We can think about that later, Bilel," she'd said whenever the question came up. But she was only a tender twenty years old, and her interlocutor was a man who always got what he wanted. In the end, she told him: "You choose for me." Then she put an end to the conversation. I felt strangely dispossessed when I gave him this choice. She may have been virtual, but Mélodie was *mine*. Of course I knew she would soon have to disappear, but she would do so with dignity and at a moment of *her* choosing. It wasn't up to Bilel. The identity change that Bilel demanded affected me. Day by day, he was psychologically killing Mélodie. She had to sacrifice everything for him: her life, her past, her mother, everybody she loved, and now this, the one thing that remained of her origins: her first name.

I checked Skype, with no intention of replying to the dozens of hearts and "ROFLs" the jihadist had sent since Mélodie had hung up. In the middle of these ridiculous emoticons, Bilel dropped this sentence: "My life, my wife, from now on, you will be called Umm Saladîne. Welcome to the true Islam." He would repeat these words the next day when we spoke again over Skype. Mélodie of course replied in the only way she could: she smiled.

The next day

André sat in on our conversations less and less. He didn't have time, and he'd already photographed Bilel and Mélodie from every imaginable angle. He also thought we had everything we needed for our story. We'd gathered a considerable amount of information on Internet jihadism, and many of our questions had been answered. He was especially convinced that our story would elicit retaliation, and he feared that the longer we let Mélodie exist, the more I was at risk. "Until we put an end to this," he said, "you're always going to want more information." I agreed with him about the risk, and the danger of retaliation, since Bilel knew my face. But I was still hungry for more. I received almost daily news from families affected by the departure of a child. They were all desperate. And I hadn't yet learned enough from Bilel to be able to help them. Bilel talked and boasted a lot, but much of what he said wasn't interesting, and it was difficult to guide him toward important topics. Still, thanks to Mélodie I'd learned things I never could have grasped on my own. But it wasn't enough. I hadn't yet come up with an exit strategy. And some part of me felt like I owed her an honorable end, after all she'd had to endure. This story went beyond professional interest; it was personal. I realized I'd put so much of myself into it that my curiosity had become both legitimate and unhealthy. André understood and let me "do my thing," but urged caution.

With André gone, Mélodie no longer communicated every day with Bilel. It drove him crazy, and I reveled in this small act of vengeance: denying him access to the woman he'd grown so attached to, Mélodie. She gave excuses: her mother wouldn't let her use the family computer, and she could only contact him when she managed to get her hands on the MacBook hidden in her room. They'd only been able to Skype twice, during which sessions he'd spoken only of marriage. I wasn't able to get him to supply any new information. I continued tracking the presence of the Islamic State's mujahideen on the Internet. Photographs depicted them proudly posing next to bodies that they had decapitated. The victims were mostly Muslims. The Islamic State uses sensationalist, Hollywood-style propaganda in its quest for expansion, convincing recruits to join its forces—and only its forces. One example: ISIS's "martyrs"* have peaceful smiles and angelic faces in propaganda photos, while the remains of its adversaries are hideously charred. In reality, ISIS takes pictures of its dead fighters immediately after they die, emphasizing their facial expressions. It lets other bodies, those of "nonbelievers," decompose in the sun before photographing them. They look devastated by the Grim Reaper. The caption is often the same: "Look at the difference: our

* *Deaths for God, according to the Islamic State.*

martyrs are happy when they meet Allah, since he is proud of them and what they have done. And look at the *kafirs*' horrible bodies. Allah is punishing them. They won't go to paradise." Guitone was especially fond of the comparison and often published these kinds of pictures online. He'd follow such posts with a picture of himself waving a Milka chocolate bar in Syria. Or, since meat was in short supply there, he'd cross the border between Turkey and Syria and, accompanied by a few of his friends, show himself sitting at a table, a Kalashnikov slung over his shoulder, a smile on his face, enjoying lamb and American soda; the caption would read: "Syria and Turkey are fighting the same battle. We're at home there. *Masha'Allah*, it's better and cheaper than in France, my brothers. You should come!" Sometimes he'd add, and he wasn't the only one to do this: "A shout-out to the DGSE if you're spying on us!"

Bilel told me similar anecdotes, but he was too high up in the hierarchy to publish that kind of proof online.

I met my gentle Milan for coffee after I left the magazine and departed with a heavy heart to head home for my meeting with Bilel. I used the commute to my apartment to clear my head before diving back into the depressing universe that awaited Mélodie. I put on earphones and turned up the volume to the Cure's "Just like Heaven," the theme song for the 1980s French TV show *Les enfants du rock*. I was too young back then to appreciate the show, but the song reminded me of my older brothers, and that memory, like Proust's madeleine, made me feel sweetly nostalgic for childhood. It soothed me for the whole journey home. The first thing I saw when I opened the door to my apartment was Mélodie's costume, ironed and hung. It almost looked alive. The cleaning lady, who comes by once a week, must have thought I'd bought a new dress.

Bilel had been harassing me with messages morning, noon, and night for the past several days. He wrote the same sentences over and over, as if Mélodie really belonged to him.

"You there?"

"You there?"

"You there?"

"You there?"

"You there?"

"You there?"

"You there?"

"You there?"

"You there?"

"Baby?"

His "you there?" questions covered dozens of pages. He badgered Mélodie on Skype, Facebook, and even her disposable phone. Meanwhile, my friends and coworkers started asking if I, the reporter, was getting too involved. I didn't understand their concern. Now I realize that I enjoyed tormenting him. For example, Mélodie wouldn't turn on the camera during their Skype conversations, or she'd trap him into talking about shameful topics. The deeper I got into the investigation, the harder it became to keep my professional distance. That had never happened to me in my career before, and I'd interviewed murderers, rapists, and pedophiles. I'd wanted to strangle them, but my face always remained neutral.

In Bilel's case—and what I'm about to say is neither polite nor journalistically ethical, but it is the best description of my "feelings"—I wanted to fuck him over. I wanted to beat him at his own game. I didn't see him as religious or even human. This murderer divided his time between taking people's lives and convincing girls like Mélodie to embrace death. I couldn't attack the powerful jihadist or his army directly, but I could attack the man's weaknesses. Namely, his thirst for recognition and domination. He thought he controlled Mélodie, but in fact the opposite was true. Bilel made me laugh when he wasn't making me sick. I believe that trust and mutual care form the foundation of love. He offered Mélodie its antithesis.

I gave the impression of having developed a kind of Stockholm syndrome, but I sensed that my friends and colleagues were skeptical as to how much of it was feigned. "Why else would you keep this up?" they asked. Because I was doing my job, and because everything I'd learned from Bilel would have taken me months to learn and understand without him. I'd told them all how he disgusted me; we'd made fun of him, but they seemed to think I was hiding something from them, as if my interest went beyond the professional. I didn't tell them everything because some part of me was embarrassed and, as I would later understand, this kind of story doesn't leave a person unscathed; I also wanted to publish this article and didn't want word to get out before I had the chance to write it. Besides, intimacy between Bilel and Mélodie never went beyond his verbal insinuations. He'd never asked to see more than her face. He didn't need to. No matter what he said, Bilel was terrifying. And again that night:

"Oh, there are you are, my wife! Were you being punished again? We need to talk. I have so much to tell you. Only good news!"

"I love good news."

"I spoke with the *qadi*[*] in Raqqa. He's looking forward to marrying us."

Stunned, I didn't know what to say.

"Doesn't that make you happy, baby?"

"I already told you that since I'm single I don't want to arrive in al-Sham without my cousin—or I have to be married first."

"The *qadi* said we can't get married on Skype."

"You wanted to get married over Skype? Is cybermarriage legally binding?"

"According to our laws it is. But the *qadi* thinks I'm too important to get married over the computer. He wants you to be on holy land. He wants us to wait until you're here to get married. He's very excited to meet you."

Meeting Bilel in Syria was out of the question. There was no way I'd let Mélodie see what life was really like in the Islamic State. All professions have their limits. I'd go there one day, but probably not disguised as a convert looking for a husband. Committing suicide would be a quicker death than that. Meanwhile, Bilel had completely erased Mélodie's cousin from the plans. When Mélodie brought him up, Bilel ignored her. He had very selective hearing.

"What are weddings like there?"

"Actually, we're already married."

** A Muslim judge.*

"Excuse me?"

"I thought I'd already spoken enough about the idea of marriage with you. I asked you to marry me awhile ago, and I talked about it with the judge, who drew up the papers. We're officially married, my wife! *Masha'Allah.*"

I don't know how I managed to maintain my composure in that moment. But I didn't have a choice, since just inches away, Bilel was scrutinizing Mélodie's face from the other side of the screen.

"I thought I told you I wanted to see you in person before saying yes. To touch your skin, smell you, have a discussion, and be able to touch your hand."

Bilel didn't say anything. Mélodie went on.

"What do you mean, we're 'officially' married?"

"As soon as you come to Syria, our marriage will be valid. As I've mentioned, we follow sharia law, and you should from now on as well. You're really my wife now."

"I'm sorry. I don't understand. All I have to do is set foot in Syria in order to become Mrs. al-Firanzi? At any time?"

If my investigation did lead me to the Syrian border, I wanted to know my exact chances of survival.

"Yeah, whenever. For as long as I'm alive, *insha'Allah*! You're really mine now."

Speechless, I blinked.

"There are just two important things to add to our marriage certificate. First, what do you want as your dowry?"

"I have a right to a dowry? Doesn't the bride's father provide the dowry? I don't have a father. Do you have money for that?"

"Of course I do, baby! I'm Tony Montana here. Except

I don't deal in drugs but in faith. ISIS is loaded. And yeah, here we respect women above all else. Here it's the man who gives his future wife a dowry, to show her that he'll take care of her for the rest of his life. So, what do you want?"

This was the first I'd heard of such an arrangement. Mélodie took a while to reply. I tried to buy time by asking other questions while I searched for inspiration in memories of past conversations with this lunatic. A strange idea came to me.

"A Kalashnikov?" Mélodie said.

Her future husband burst out laughing. I didn't know how to interpret it.

"That's what you want? That's it? I'm proud of you, but you know you could have asked for much more."

"I could have? Like what?"

"I don't know, a palace, a castle, some pretty horses. Or the life of someone who's offended you."

"That's okay. All I want is a Kalashnikov."

"In any case, the emir of Raqqa, a very important man, has already found a beautiful big apartment for us."

I had trouble imagining a two-bedroom apartment in Raqqa.

"That's really nice. What's the apartment like?"

Bilel's face fell like it always did when he lied. He lowered his eyes and scratched his slightly tilted head. I was as familiar with this posture as with his dreamy looks. What an actor. His exaggerated expressions were becoming increasingly irritating with each passing day.

"Well, it's big . . . and it's nice . . . you'll see! You'll have to decorate it. Okay, I have one more question for you, and it's really important."

"What is it?"

"I want you to promise me first that you'll give me an honest answer, because we take this kind of thing really seriously."

"I promise. Ask."

"Are you a virgin?"

"Yes."

"Really? Because the *qadi* is waiting for your answer so he can include it on our marriage certificate."

"Oh? Because my virginity is all of Raqqa's business?"

"Of course not! It only concerns your future husband and the supreme authority. That's it. You crack me up. Mél-odie, you're so sweet and pure."

Personally, I didn't find any of this funny. Bilel went on.

"You know, lying about that is punishable by death. There will be women to check if you're pure on our wedding night."

I forced a sour laugh.

"Don't disappoint me. I've already told everybody you're coming, including the other brothers and the border police. I've really put myself out there for you, so don't make a fool of me. Be strong; come to Syria. You're a real lioness, my wife."

"The border police? What is that? A real police force or a friendly arrangement?"

This was an allusion to Turkey, which has been accused of turning a blind eye on border passages.

"Both . . . I'll explain more once you're on your way. It's too risky now. Cops and journalists are everywhere. They're all *kafir*s and deserve to die."

. . . .

Mélodie let out a nervous laugh and changed the subject. Bilel had promised to help with her jihad, but aside from saying she'd have to choose between Holland and Germany, he hadn't provided any details. Since his hearing was selective, Mélodie played along with his plans. Yes, she would do her jihad. She'd go through Amsterdam. I already knew the route: from Amsterdam she'd fly to Istanbul, and from there to Urfa or Kilis. Mélodie, however, wondered: what next, after Holland?

"By the way," she said, "we have to talk about the trip. And about Yasmine."

"Who's Yasmine?"

Bilel had already forgotten about the fifteen-year-old girl for whom he'd promised "a good husband and a dream life." I was furious, which bled into my avatar's reply.

"Are you kidding me? Yasmine, one of my best friends, you promised to help her! We've talked about it several times. I asked if her age was a problem, and you said no, and you explained sharia. . . . I'm not going without her!"

"Oh, right. The minor. Fifteen, you said?" He rubbed his chin, his eyes bright. "Calm down, baby! You shouldn't get mad at your husband, it's *haram*. We'll take care of your friend. Don't worry."

"Who's 'we'? I've heard stories about women who arrive in Syria and are treated like sex slaves for your fighters."

"Don't listen to what those idiot Frenchwomen have to say. They're not here; I am. I'm your husband, and you have to listen to me and only me from now on. Do you understand? Anyway, I promised to help your friend Yasmine, and I will. The day before you leave for Holland, I'll explain what to do. Don't worry, I or someone important will meet you in Istanbul."

"Okay, I understand. We have to go by ourselves all the way to Istanbul?"

"Yes, but we'll get into that when the time comes. *Masha'Allah*. The main thing is that you get here fast. Especially you, Mélodie."

"I hear my mom in the stairwell! I have to hang up and go to my room."

I was exhausted. Talking to Bilel was becoming increasingly tiresome. I regretted not keeping more distance. I'd intended on using him to confirm or deny specific facts, and to learn a bit about his frame of mind. I hadn't planned on getting tangled up in this madness.

"Okay, my wife. I love you for the sake of Allah."

"Okay. Don't forget, after tomorrow I'm going to Tunisia for a week to take an Arabic class with some other sisters. I won't have Internet access."

"Forget that. You'll learn the language faster when you're here."

"Well, it's already planned. It'll be good for me before my jihad. My mom is coming; I have to go."

"Did you hear me say I love you?"

"Yeah, but I really have to hang up now. *Masha'Allah*."

I finally disconnected from Skype and collapsed on my sofa, worn-out. I thought of Yasmine. I'd given her the same name as one of my friends, although the two girls had nothing in common. Absolutely nothing. I would have loved to be with her right then, eating sushi and watching something stupid on TV. Instead I had to come up with a viable backstory for Mélodie's Yasmine. I dug through my memory, thinking of all the girls I'd met over the years, in particular Wendy, Marlène, and Charlène. My "bad girls" from the ghetto suburbs of Lyon. I met them six years ago, and we're still in touch. They did some stupid things in their past, but they've changed their lives and today they're on the straight and narrow.

Anyway, the big news of the day: I was married! For the time being, I ignored much of what Bilel had said to Mélodie that night, particularly the points that might thwart my job as a journalist, downplaying some of the serious implications of our conversation. Before going to sleep, I'd call Céline or Andrew, my trusted friends, and take a bath, hopefully erasing the image of women inspecting my virginity.

But when I walked into the bathroom, I saw Mélodie in the mirror. I'd forgotten to remove her veil again, and suddenly I was too tired to do anything. All I wanted was sleep.

Thursday

I was having lunch on a restaurant terrace with Lou and three other friends. Our mood matched the spring day's sunny weather. We ate cheeseburgers; we laughed and joked. All of a sudden, Mélodie's disposable phone rang. My friends and I stared at one another, instantly aware of who was calling. I answered, stepping away from my friends so they wouldn't see me playing my role. I didn't want them to worry, and if I'm honest, I was afraid they'd judge me. Bilel had been relentlessly contacting Mélodie since his marriage announcement. He sought reassurance and tender words. He was also becoming increasingly anxious to know exactly when Mélodie would arrive in Syria.

Our phone conversation at lunch was brief. Mélodie told him what he wanted to hear. But, without quite realizing it, I must have been giving my friends an alarmed signal. A few minutes later, I calmly rejoined the table. As far as I was concerned, there was nothing to get worked up about. I didn't yet understand how schizophrenic this story was making me. I was sure of myself. I thought I'd become a machine that could adapt to anything, and even suppress my feelings of disgust.

Besides, the story was nearing its end. We'd planned to plot the investigation's denouement that afternoon. I wolfed down a mouthful of fries, and when I looked up, all eyes were on me. Bilel's call had apparently spoiled everyone's appetite. My friends were pale-faced, terrified that Bilel might be able to trace my location via Mélodie's cell phone. "Have you thought of what would happen if he came here?" Of course I had. Only, I reminded them, the phone wasn't associated with anyone, and I'd scrambled my IP address prior to engaging in any conversations with him over Skype and Facebook. They didn't seem convinced. And though I was slowly able to change the subject, the light mood was ruined.

We went back to work. During our smoking breaks, I sensed my friends' tension and stress. I felt guilty and tried to reassure them. I hated myself for worrying them. That night, they sent text messages filled with concerned advice and support. Comforting emoticons accompanied their serious words. I understood without understanding. Of course I couldn't plan for everything, but I'd taken precautions. Besides, I was surrounded by seasoned reporters who had worked on more dangerous missions than this one. Strangely, people seemed less worried about them right now. Basically, although my friends' concern touched me beyond words, I didn't think it was entirely justified.

Mélodie's conversations with Bilel were now rapid exchanges, and she controlled the tempo. She tried to avoid the terrorist's saccharine expressions of affection and redirected the conversation to the issue of her trip, an important step for all would-be jihadists. There was also the question of

Yasmine, a minor. Although it involved having to come up with thorny explanations for her future husband, Mélodie was unwavering in her resolve to bring her friend. After work, I allowed myself a quick drink with Hadrien, before rushing to meet André, who, now that the end of the investigation was planned, would be working with me again. We just had a few details to iron out in our strategy. Bilel had asked Mélodie to marry him. He assumed we were bound to each other in life and death, but meeting him in person was never up for discussion, for the obvious reason that going to Syria these days amounts to a suicide mission, especially for Europeans. And especially for a French journalist. Reporters take risks—it's their job—and those leaving for Syria know that their return is not up to them. I at least don't have any children. Still, we needed Mélodie to take this trip in order to wrap up the story, so we came up with a solution.

When I got home, I opened my computer and quickly checked Bilel's flood of amorous messages as I threw on my veil and the rest of my costume. That night, Mélodie didn't have any time to waste. Lou and I were leaving for Tunisia in the morning (Bilel thought Mélodie was taking an intensive Arabic class in a religious school) and I wanted to enjoy an evening with Milan before my trip. So Mélodie claimed her mother would be returning home at any minute; they'd have to speak quickly about her itinerary to Syria. He'd already explained that she'd have to "leave everything behind, cut all ties"; she couldn't even leave a letter for her mother: "just disappear; you can send news once you're here—and only then." But he didn't offer any concrete details, and I needed them to plan my trip with André.

After some uninteresting conversation, Bilel finally said, "Like I already said, you'll take a flight to Holland or Germany, whichever you prefer. When you get there, turn off your phone and get rid of it. Buy a new prepaid phone and get in touch with me via Skype so I can be sure it's you. Then I'll give you instructions on how to get to Istanbul and everything else."

"No, Bilel. You have to tell me more. This isn't just any trip. Consider Yasmine. She's younger than me, and she's flipping out."

"Well, make sure that on the day of your departure, your mom thinks you're sleeping over at Yasmine's house and vice versa. Leave early in the morning. Take a bag that's about the same size as usual. Yasmine should do the same. Go to the airport. Be careful not to get caught; cops are everywhere. Act normal. Don't look scared, and don't turn back. Your home is here; that's all that matters. Okay? Are you a lioness or aren't you, my wife? If anybody asks you any questions when you arrive in Istanbul, say you're with Doctors Without Borders." He winked. "Don't forget your passport! It's really important in Syria."

In reality, I knew they confiscated passports from Europeans when they arrived, in case the new recruit tried to escape.

"I'll tell you more over a secure line once you've completed the first leg of your trip," the terrorist continued. "I'll give you the number of the mother who will be meeting you in Istanbul."

A "mother" . . .

"Do you promise it'll be a woman? You don't want men

to look at women, so you wouldn't let one of your brothers see me before you do, right?"

"Of course I wouldn't! Never! Are you crazy? Besides, Turkey is home. We do what we want there. I'll be there when you get to the Syrian border. *Insha'Allah*."

"Where exactly?"

"I'll let you know. Unlike most of our members, I can't set foot in Turkey. Other places, yes, but not there."

"In Iraq?"

"Yes, my wife. But shh. We have a secret mission to take back Iraq. *Insha'Allah*, I hope one day we can live there together."

"I thought mujahideen could visit Turkey's border towns whenever they wanted. At least that's how it seems from their Facebook posts. Why can't you?"

"Right, well, for me it's different. I'll explain later on another line. By the way, now that we're husband and wife I want you to close your Facebook account."

"Why? I haven't posted any pictures of myself, and my posts only encourage others to do good in the world."

"Because I said so. You're my wife now, and a good wife does whatever her husband asks."

Fine. Mélodie agreed, but I wasn't going to close the account. Not yet. For a self-proclaimed "terrorism expert," Bilel behaved strangely. On the one hand, he insisted I keep unimportant details secret, while on the other hand, he made offhand comments about how his army was going to conquer Iraq.

"Okay, I have to go. I'll talk to Yasmine and make the necessary arrangements. I'll keep you posted."

"Do you have money for the tickets?"

"Yasmine doesn't, but I can work it out on my end. It'll be more difficult once we get to Syria."

"Don't worry about that, baby. I have everything you need and more here. I already told you: you're my jewel, and Raqqa is your palace. You'll be treated like a princess."

Mélodie cut short the conversation. She skillfully repeated "I love you for the sake of Allah" a few times and hung up.

My story was finally taking shape. I'd learned a lot, but it wasn't over. I decided Mélodie would go through Amsterdam. I had a contact there I wanted to interview as part of my investigation. It would be difficult to see young people traveling alone and not wonder if they were candidates for jihad. Soon it would be all over. I still didn't know the final steps of the journey, after the new recruit has left his or her country of origin. I fell asleep thinking of verses and precepts in the Koran that have nothing—or very little—to do with what Bilel had taught me. The word "mother" appears thirty-two times in the Koran. I thought of the "mother" who was supposed to meet two young girls in Istanbul and lead them to a group of assassins.

Friday

At last I met Lou at the airport. We'd both been looking forward to this vacation. On the flight, we ate M&M's, read gossip magazines, and took selfies. Two ordinary girlfriends. "Ordinary life" had become so distant, and this trip was like a breath of fresh air before the last phase of my investigation, which would take place upon my return. I'd nevertheless packed Mélodie's equipment, just in case.

We were waiting to pick up our bags when Mélodie's phone, buried deep in my purse, began to vibrate. Bilel had left several text messages, and there was one from a girl I didn't know named Vanessa.* "Hi, sister, I'm six months pregnant and I'm going to Syria in a few days. My husband fights in your husband's battalion, and he gave me your number to see if we could travel together. Given my condition, I'm afraid to travel alone, and I'm scared my stomach might draw attention. *Subhan'Allah*†, my sister. We'll be there soon." She gave me her Skype name so we could speak

---

** Her name has been changed.*

† Translator's note: "Glory be to God."

online. At the time, I didn't pay much attention to the message, suspecting it wasn't real. A girl who came out of nowhere, who told me she was very pregnant, and who wanted to travel only with me . . . It was strange. Was Bilel testing me? I didn't care. This vacation was my time. I wouldn't spend it thinking about Mélodie, and especially not Bilel. I tossed the phone back into my bag.

When we arrived in our room, the bath towels forming two intertwined swans surrounded by red rose petals at the foot of our bed sent us into a fit of hysterical laughter. It was eighty degrees outside. I opened the window to feel the sun on my face. Before going down to the pool, we unpacked our clothes: short shorts, tank tops, skirts, sneakers, and sunglasses. We'd both brought sweaters, and when I took mine out, Lou saw Mélodie's djellaba. Sniggering, she picked it up between her fingers and peered at me questioningly. I said I'd thrown it in my suitcase at the last minute and didn't plan on unpacking it or the *hijab*. While I put on a light dress over my bathing suit, she teased me, trying to imagine what I looked like in the veil. "Do you want me to put it on?" I asked, laughing. She raced across the room like a child, crying, "No, no, no!" Lou had already seen pictures of Mélodie; she didn't want to see her in real life. But we were on vacation, and I was in a provocative and jokey mood. When she came into the bathroom, I was primping myself, in my veil. She covered her eyes and cackled, then she took a picture of me while I put on sunscreen. Of all the photos taken of me at that time, this shot is the only one that doesn't leave a bitter taste in my mouth.

We spent our vacation lazing about, mostly sunbathing by the pool and engaging in "girl talk." Then, one afternoon, the telephone rang. It was Mélodie's cell, which I had more or less unconsciously forgotten to leave in the room. We looked at each other as if a ghost were calling. I took some distance from my friend and answered. Bilel was beside himself with worry. His tone wasn't the least bit threatening; rather, he sounded like a little boy. He hadn't heard from Mélodie in seventy-one hours. Was she okay? Had she forgotten her husband? She reassured him in a low voice, but the reception was bad and Bilel couldn't hear what she was saying very well. I found myself just feet away from the other vacationers, my skin slick with oil, pacing around in a pink bikini and talking in a relatively loud voice:

"*Masha'Allah*, baby. Of course I'm still yours, but I don't have Internet access, and I spend a lot of time studying. We're not allowed to talk on the phone during class."

"But you're only with other sisters, right? And you're wearing your *hijab*? I tremble at the idea that others can see you."

If he could see me now . . .

"*Bismillah*, it's just women here, and we're all covered from head to toe. We're only interested in learning."

The situation was absurd. Hearing what I was saying, the other guests stared at me from their lounge chairs. I felt like I was putting on a comedy routine. Wrapped in her beach towel, Lou was choking with laughter. Seeing me half naked and speaking a language somewhere between French and a very crude Arabic was too funny. I turned around to avoid their gazes. I was beet red, and I didn't know where to stand.

"I'm so glad to hear that, my wife. I was afraid they'd done something to you, or worse, they'd turned you against me."

"No, of course not."

"Then why haven't I heard from you?"

Because I was on vacation from you, Bilel! And since he never listened to anything Mélodie told him, she reminded him.

"I told you I wouldn't have Internet access. I'm in the middle of nowhere in Tunisia, the connection is bad on my cell phone, and it's expensive."

"Okay, my wife. It hurts me to think of you there. Do you have money, at least? Soon I'll take care of you and our little family."

"I'm fine, but I don't have many minutes on my plan, and I'm trying to save them in case of an emergency. How about you? How are you?"

"I'm fine. Don't worry about me. I have news."

"What news?"

"I'll tell you later. But everything is working out for us . . . *insha'Allah*. When do you arrive?"

Bilel was obliquely referencing attacks on some big Iraqi cities, which ISIS would conquer in June, as well as to the caliphate that would soon be established.

"Soon."

"When?"

"I have to see with Yasmine and find the cheapest date to travel to Istanbul."

"But when? When? Mélodie, my wife?"

It had only been a few minutes, and I was overcome with

stress. The weight of it all came back: the exhaustion, the feeling of oppression, the lies, my sofa, which will always remind me of our conversations.

"Probably next week."

"Okay. Keep me posted, all right?"

"Of course."

"Promise!"

"I promise, Bilel."

It was true. I was really going to Istanbul, as per his instructions, but André—not Yasmine—would accompany me. The plan was simple: Bilel had told me a mature woman was to meet us there. She'd be expecting two young girls dressed in burqas, and she wouldn't notice two friends in jeans and sneakers. We'd blend with the crowd and quickly disappear into a taxi. André had experience taking pictures "on the sly" and would capture the "mother" on film. I would walk a few steps ahead of him in order to point out the woman if need be. While the recruiter looked for Yasmine and Mélodie, André and I would continue on to Kilis, a city near the Syrian border. Kurds controlled it, and we figured it would be safer than other places.

Kilis hasn't been ravaged by the war, but it has seen misery. We had an interview scheduled with Guitone there. I wanted to question him on some of the things Bilel, Abu Mustapha, and others had told me. To compare answers. The story would end there, with a photograph of Mélodie looking out at the Syrian border from behind. The journalist would stop at the doors to hell, and Mélodie would step through them. I'd finish the story without putting anyone in harm's way. Not physically, at least. The end was in sight, giving me the motivation I needed to let go of the investigation—and of Mélodie. Enough splitting myself in two. Enough encroachment on my personal life. *Khalas*. We were finally wrapping this up. At least that's what I thought as I hung up the phone with Bilel in Tunisia.

Four days later

Our short break in Tunisia was coming to a close. The day before our departure, Bilel pestered Mélodie, inundating her with messages. He was frustrated with the briefness of our poolside call. He missed her so much it hurt. His nights were less enjoyable when he couldn't see her face in the evenings, and his days were almost insufferable. He'd been asking to see her for days. Why did she deny him? His tone hardened. His diligent efforts to brainwash Mélodie, along with their long hours of conversation, had affected him. He was a man in love. And that was very bad news for me. Bilel had many faults, one of which was his outsize ego. If he found out that the girl he loved was a journalist, I would be in serious danger. I hoped I was wrong. A betrayed man can become an entirely different person, and I didn't want to imagine what that might mean for a jihadist like Bilel. I knew he wasn't magnanimous enough to protect the woman he loved from his hate and desire for revenge.

In our hotel, the only area with Internet reception was the lobby. I'd spent the past several days like the other vacationers, hanging around the pool in my bathing suit and giggling with my girlfriend. I was uncomfortable at the idea of wearing my *hijab* in the lobby and reassuring my cyberdouble's husband. And I could forget about the long black dress. . . . At the same time, I knew Bilel had to see me; otherwise he might become suspicious. I'd position my computer's webcam so Bilel would see Mélodie in a close-up shot, and I crossed my fingers he wouldn't ask me to move the camera. As I set up for the Skype conversation, I sensed Lou's stress behind her otherwise relaxed facial features. Was she mad at me for doing this? Or was she scared for me? Lou is an extremely private person and hates all forms of intrusion. That night, Bilel would intrude on our vacation. He'd already dampened a few carefree moments that week; now he was ruining our last night.

We sat down on one of the hotel lobby's generous circular benches, our MacBooks resting on our laps. I told Lou I could do this by myself, but she also needed to use the Internet. Maybe she wanted to remain close by, to protect me. I drew my black veil over my short white dress, my feet in flip-flops. My face was very tan for someone who had supposedly been inside studying. Lou pretended not to see me, which was just as well. Now wasn't the time to joke around. This wouldn't be easy. I thought of Lou as a little sister, and it bothered me to play this part in front of her. Still, she knew what I was doing. Everything would be all right, and the conversation would be short; Mélodie didn't have much time.

Bilel finally saw Mélodie's face. His eyes were filled with worry.

"How are you? I can't continue my quest without you. Don't do that to me ever again, my love, my life, my wife."

"*Masha'Allah*, Bilel. I'm sorry to have worried you, but Internet access is scarce here. And I have to be discreet. Our Arabic teachers have been warning us about the dangers of jihad, and I don't want to awaken their suspicions."

"Don't listen to them! Your place is here, with your husband. *Insha'Allah.* I missed you so much. I'll be able to sleep a few hours tonight."

"But you've slept a little since I've been unavailable, haven't you?"

"Not really. I told you when we first met that I was fiercely devoted to Allah. Before I only had work, but you've given me another reason to live."

I hung up. For so many reasons. First, I'd been watching Lou from the corner of my eye. She was also having a conversation on Skype. I was curious if she'd heard what Bilel had just said. It didn't seem like it. She was still pretending to ignore me. Then, I don't know why, but the hotel's activities staff was always hovering around us. There was one in particular, whom I'd nicknamed Mosquito. They never talked to us, except when they were playing water polo next to our lounge chairs. Sometimes they enjoyed making the other vacationers laugh by waking us up from our naps. For some reason they chose that night to hit on us. If Bilel heard even the whisper of a male voice, it would be a disaster. There was no doubt he'd call back soon. Everyone in the lobby was staring at me, surprised and even disapproving. Bilel's profile picture reappeared on my computer. I clicked "Answer."

"Tell me you love me, Mélodie!"

"I can't hear you, baby, but I wanted to say that I'm

doing well, and that I'll be back in France tomorrow. We'll talk when I'm in Toulouse again. I promise I'll have time."

"My wife, I hear you and I'm reassured, but I hate that other people can see you."

"But I'm wearing my veil."

"It doesn't matter, you shouldn't go out like that. It really hurts me."

"But I'm being good, and you said that if I covered myself, I could go out."

"I meant here!* Not in depraved countries!"

"I'm in Tunisia."

"That's worse. Their women wear high heels! Most of the jihadists in al-Sham come from Tunisia. They come here because they're disgusted by the wolves transforming sheep into easy women and female infidels. I swear before Allah that I'll kill anybody who comes near you!"

"No man has spoken to me; don't worry . . . okay?"

"Okay. But I don't want anybody putting ideas in your head. Focus on improving your Arabic, that's it. You look different—tanner."

I was expecting him to notice, and worried he might start to suspect something, but I didn't perceive any doubt in his voice.

"Different? Better, I hope. A little color is pretty on women."

"You look like the girls I know here," he said, his voice bitter. "I prefer lighter skin and the complexion Allah gave

---

* *In Raqqa.*

you. You look less innocent when you're tan. But it brings out the green in your eyes, which I'll remember as I go to sleep."

I breathed an internal sigh of relief. Lou kept peeking discreetly at my screen. I logged off a few times, pretending the connection was bad. Still, Mélodie had a difficult time getting rid of her "husband."

When she finally succeeded, I quietly took off my *hijab* and turned toward my friend. I recognized the look she was giving me. She was annoyed. Among the turmoil of emotions, it was all she could express. I managed to coax a few words out of her. She explained that hearing my conversation with Bilel had been disturbing. So she *had* been listening. . . . She thought it was too dangerous for a journalist to expose herself like that. There was a real intimacy. She'd sensed Bilel's feelings of ownership toward Mélodie and it had been chilling.

Lou said she'd catch up with me later. She had more Skyping to do. When she came back up to the room, I was in bed. She silently lay down. I waited, and after a while, she smiled. We didn't talk about what had happened. I took out my heavy artillery of nail polish, and I painted her nails red.

Wednesday night

There was my couch. As dark as Mélodie's clothing. Usually I love returning home after a trip. It's a bubble whose protective force seems to strengthen while I'm away. When I'm in my apartment, I feel like nothing can touch me. I especially love coming home to my big dog. I adore him. He's like a giant stuffed animal who sleeps all day. I named him after a Nancy Sinatra song. That night, after my shower, I buried myself under the covers. I was going to call Bilel from my bedroom. It didn't happen very often. André didn't like it, but sometimes, due to time constraints or lighting issues, we didn't have a choice. The atmosphere in my bedroom was soft and cozy, the glowing light mostly from candles. I positioned my computer on my lap and pulled on just the veil. The camera angle was such that Bilel would only be able to see my face. I'd made a cup of hot tea, which I set on my wooden night table. I called. The sound of his voice exhausted me. I couldn't stand this man.

"Did you miss me? Do you love me? I love you so much, Mélodie, my wife."

"I can't hear you very well. How are you?"

"I have something really important to tell you! I gave your number to one of the brother's wives. She's supposed to arrive soon, but she's six months pregnant, and she would be relieved if she could travel with you, since you're the wife of an emir. She can also give you a lot of advice. Her name is Vanessa."

Panicked, I hung up. The message I'd received in Tunisia had been real after all! A very pregnant girl, who was obviously a minor, wanted to travel to hell on earth. Bilel called back, and I slammed my laptop shut, slightly damaging it in the process. I looked for Vanessa's text message in Mélodie's phone. I wanted to reply right away. When I found it, I quickly typed a message.

"*Salaam alaikum*, sister. Sorry not to reply sooner. I needed to think about our trip. Is this really a good time? Especially given your condition?"

Vanessa replied immediately in a mix of Arabic and French. Her message wasn't entirely clear, but I understood that she was determined to join the father of her child. She begged me to travel with her, promising to answer any questions I might have.

Panicked, I didn't know what to do. I wondered if I should call the police. I'm a journalist. I don't turn people in to the authorities, especially not fragile young women. But would it really be snitching? Or would I be protecting her? Stressed, I weighed the possibilities. Mélodie ended up suggesting they wait another week so she could better prepare for the trip. It was a way of buying time. Vanessa warmly agreed, which ended our twenty-minute text message exchange. It gave me more time to think and discuss the situation with my editors. I had to call Bilel back. Even more reluctant than before, I logged on again to Skype. Mélodie mumbled an excuse concerning her sister, then let him talk. I had gathered a large amount of information, more than I'd hoped for. All I wanted now was one thing: a fast ending.

I didn't listen to his sweet nothings. Mélodie simply reminded him that she was leaving to meet him in two days. I wanted to make sure he was listening. I lifted my cup of tea, which was now cold, and blew on it reflexively.

"What are you doing?! Stop, that's *haram*! Mélodie, stop!"

Stop what? I didn't understand. His eyes were popping out of his head. Why was he so excited? What had I done?

"Don't blow on your tea!"

"But it's hot!"

"It's *makloum*!* It's really bad. Do you know why?"

"No, why?"

"Because it doesn't adhere to Islamic law, but to the laws of your country. Come on, Mélodie!"

Right! Because it's common for countries to include a line about "behavior toward hot beverages" in their civil codes.

"I still don't get it."

"It's written that you should not change the nature of things. Sharia has strict laws: if something happens to you tomorrow—like if you're assaulted or mugged—you're considered a *kafir* if you file a complaint in your country. You would become my enemy and Allah's enemy, because turning

* *Forbidden.*

to men for justice makes you an infidel. For instance, does your mother have insurance?"

"Of course she does. She has several kinds and even for loans."

"Well, that makes your mother an enemy. She doesn't respect our laws, which means she doesn't respect Islam. Hence, you don't owe her anything. I suggest you review your lessons on the *tawhid* and the *dawa*,* and turn your back on enemies."

Great. I was dying to slog through the Old and New Testaments, the Koran, and all the laws that make up sharia. The more I talked to this man, the more I loathed him. Bilel reminded me of a pedophile when he spoke of Yasmine with stars in his eyes. When he talked to Mélodie, all I saw were vice and lies. He was not, as I'd first believed, a wolf covered in sheep's clothing. He was the devil. I'd seen the look in his eyes when he'd screamed at Mélodie not to blow on her tea. Or when he'd said her mother was an enemy. I was still interested in studying the Islamic State, al-Qaeda, and everything that was going on in the Middle East, but without Bilel—an evil genie who popped out of his lamp whenever he pleased.

Mélodie was getting ready to hang up when suddenly Bilel asked, "How old are you again?"

"You've already forgotten?"

"No, but I'm unsure."

"I'm turning eighteen soon," I said, to see his reaction.

"Oh, okay, darling. You're so beautiful."

** An invitation to non-Muslims to listen to the message of Islam.*

"I was joking! I'm twenty. Yasmine is the minor!"

"*Masha'Allah*, my wife. You shouldn't play with me like that, my love."

"I know that you're thirty-eight, and your birthday is January eighth. It says so on your Skype profile."

"That's just to throw people off."

"You're not thirty-eight?"

"I am, but I wasn't born in January. My birthday is June 6, 1976. I've had to change my identity a few times."

My heart skipped a beat. Any other journalist would have continued questioning him after that admission, but I couldn't. I'd had enough. In that moment, I wanted to tear off my veil. Instead, I replied in my real voice, incapable of pretending to be someone else.

"Bilel, I have to go."

"Already?"

"Yeah. Good night."

"But—"

I slammed my computer shut, cutting him off for the second time that night. After Milan had seen me playing Mélodie, this was the second conversation that had affected me personally. The second time, and there wouldn't be a third, that he had inadvertently intruded on the most private part of my life. I lit a cigarette and fought the urge to throw my Mac across the room. A long time ago, one of my older brothers, the one who was born the same year as Bilel, said something I've never forgotten. My family had just moved into a bigger apartment overlooking a cemetery. My brother and I stood by the window and smoked Marlboro Reds, silently contemplating the vast and desolate expanse. At one point my brother thought out aloud: "Every omen has its destiny." Not long after, he died from no longer wanting to live. He'd just turned twenty-six on the sixth of June. Bilel and my big brother were born on the same day in the same year. One was still alive, the other wasn't.

I've never told anyone that before. Not a day goes by that I don't think of my brother, with his dark eyes and long lashes. I thought of my parents, who still live in the same apartment, and I took this horrible coincidence as a very bad omen.

Early the next morning

The sound of the phone ringing tore me from sleep. Still half dreaming, I answered. It was André. His father had just died, and he couldn't come with me to Amsterdam the next day. He apologized for having to withdraw from our investigation. His voice was a mix of pain and disappointment at not being able to be with me after everything we'd been through. Of course he needed to be with his family. Tears welled up in my eyes. I didn't care about the story; I just wanted to give him a hug. I didn't know what to say to reassure him. He should forget work, Bilel, and everything else. He asked me to tell the magazine, which I did immediately. I took a shower, got dressed, and rushed out of the house. When I arrived at the magazine, my editors were already hustling to find a "trustworthy" photographer who could stand in for André at a moment's notice. Somebody with enough experience to be able to watch over me when we were at the border with Syria, and who was familiar with the mentality of fanatics. Given the risks involved, it would have to be somebody who could keep his or her cool in dangerous situations.

The editorial staff was extremely cautious—even to the point of wavering over whether or not to send me—since Europeans who travel to that part of the Middle East have few assurances they won't be kidnapped. Two members of the senior staff, the head of photography, Hadrien (everybody who'd been supervising this story since its inception), and I spent hours narrowing our choices down to two candidates. Between those who refused to go to that part of the world and those whom the photography department deemed unqualified for the mission, there were very few options. In the end, we decided on Charly, an excellent photographer whom I only knew by reputation and some of the striking pictures he'd taken of various conflicts and crises over the past thirty years. The journalist in me was reassured. Everything would be fine with Charly at my side. But on a personal level, I would have preferred someone just as qualified but whom I knew better, someone like Julien. Charly was going to see me playing a part and flirting with Bilel, and that would be his first impression of me. The actress in me, the one I'd been cultivating these past several weeks, would have preferred it to be someone who already knew me. The few times I'd seen Charly, he'd struck me as a serious person. I hoped he wouldn't judge me. The editorial staff briefed Charly, and we spoke later that night so I could provide more details. I'd barely gotten to say a word when he interrupted me:

"Please, I don't want to know your real name. As far as I'm concerned, you're Mélodie. I don't want to risk calling you by your real first name. You can tell me after we've finished the job."

I smiled. Charly wasn't uptight at all. And I thanked God, if he existed, for putting him on my path. After that, we had a long talk. Charly found himself thrown into a complex story about which he'd known absolutely nothing twenty-four hours before, and which would lead him to Holland in the early morning and then to the border between Turkey and Syria a couple of days after that. It wasn't a common scenario, particularly since the journalist was also the protagonist. He wondered how he would manage to capture my image without giving away my identity. I sensed he was uneasy photographing a conversation; he was used to hiding under tanks and working while bullets flew overhead.

The magazine had given him three goals. One: "snap" a covert shot of Lola,* the girl I was meeting in Amsterdam. Two: get pictures of the "mother" who was supposed to greet Mélodie and Yasmine in Istanbul. Three: take pictures of Guitone, and then of Mélodie in Kilis. Lola was the reason we were stopping in Holland. She'd contacted me out of the blue, wanting to tell her story, and she happened to live in one of the cities Bilel had designated as a stopping point. It was convenient, and I loved Amsterdam. Hadrien would also be there that weekend to attend World Press, the most prestigious press photography contest in the world. It's basically Cannes for photojournalists.

Among her online encounters with other sisters, Mélodie had met the Lola in question. The girl had almost gone to Syria to be with a man she thought she loved, an ISIS fighter. Happily, the authorities had stopped her a few minutes before

---

** Her name has been changed.*

her departure for Syria. I was dying to know her story, which seemed very similar to Mélodie's experience. After Lola was caught, the police banned her from all contact with her jihadist. Her anger faded, and she saw her supposed "family" commit countless acts of violence. She examined dozens of Western and Middle Eastern websites and was able to form her own opinions. She realized she didn't agree with ISIS's ideology. Although she now understood the perils of living in a war zone, she was still devout and lived in a home for girls. Her parents were Protestant and wouldn't allow her to live with them unless she agreed to remove her burqa and gloves. Lola refused. The authorities kept tabs on Lola, so it was safer for us to meet in person. She was scared that giving an interview to a journalist would anger ISIS or prejudice the police against her. The aim of this trip, in addition to Mélodie's mission, was to hear Lola's testimony. The interview would not only feed my investigation; it would provide Mélodie with plausible explanations for Bilel when she told him she couldn't join him after all.

It wouldn't be a simple task. The fighter wouldn't easily allow his prey to escape. My interview with Lola was scheduled for 4 p.m. From the other riverbank, Charly (aided by Hadrien) would use a telephoto lens to take pictures attesting to the veracity of our encounter. He would frame the shots around our profiles so that we—especially Lola—couldn't be recognized. Then we would rush back to the hotel room for Mélodie's final conversation with Bilel. I was impatient to learn about the last leg of her journey. Then, the following morning, while Bilel thought I was on a plane for Istanbul with Yasmine, Charly and I would have already landed.

If everything went according to plan, after Charly took the "mother's" picture, we would have plenty of time to catch our connection to Kilis. In addition to the scheduled interview with Guitone, I wanted to see for myself the atmosphere and desolation in the border town. I hoped to speak with men, perhaps some women, and even teenagers. I wanted to hear what they felt as their dreams of joining the Islamic State became reality. I needed to see the demarcation line. To take a deep breath and find myself on the right side of the mirror. To free Mélodie, and most of all, to free myself.

I would probably go to one of the hotels known for hosting guests for a single night before their crossing to "the other side." Once there, Mélodie would send Bilel a final message telling him she had to turn back; the authorities hadn't let her board the plane for Turkey. She'd say she was being watched and had to return to France for the time being. Then she would disappear—vanish. And Bilel would never hear from her again. My story on "digital jihadism" would end in Kilis, with a remark on how few obstacles got in the way of those wishing to embark on such a terrifying journey.

But nothing went as planned.

Friday the 25th

I caught sight of Charly in the crowd of passengers. "How do you know what I look like, Mélodie? We don't know each other." I'd told him his reputation preceded him, mine less so. . . . We immediately got along. To start, he kept making fun of Mélodie, and I love it when people make me laugh. Charly was not at all how I'd imagined him. He and André were really different. Charly was levelheaded, spoke softly, and made free use of his dry wit. He was always in control, even if it didn't seem like it. Humor and sangfroid were his weapons. He downplayed stressful situations and always completed his missions. I'd brought my old Rolleiflex camera, a gift from my dad. It must have been from the forties. I didn't use it often. "Look, it's easy," Charly said, taking it from my hands. We were in the middle of a lesson when Hadrien arrived. He and Charly had known each other for years. In the end, although I missed André and felt a dark premonition, we boarded in a good mood, all impatient to arrive.

I started typing my introduction on the plane, since from now on everything was going to happen really fast. With less than an hour before landing, I listened to "Eye of the Tiger," the theme song from *Rocky*, in one ear. I got really into it. If the flight attendant didn't give me a packet of celery salt with my tomato juice, I'd hit her with an uppercut—the one I wanted to give Bilel.

Everything sped up when we reached Holland. Lola pushed back our meeting by a half hour; she seemed to be wavering. Then she called again to delay the interview another hour or two, claiming she couldn't leave the home where she was living whenever she wanted. Shit! Charly and I both knew that it wasn't a good sign when a contact continually changed a meeting time. We crossed our fingers that she wouldn't stand us up altogether. While I went to buy an untraceable prepaid phone (as Bilel had instructed), Charly and Hadrien braved the sweltering heat to look for the best places to take pictures. It was late April and almost eighty-five degrees. We also happened to be there in the middle of Koningsdag, the Dutch national holiday. DJs played deafening music from every street corner. Residents were dressed in traditional orange. They laughed, drank, and sang. Many wore wigs and were carrying retro stereos on their shoulders, like Run-DMC. The mix of heat and noise, together with our tight schedule, started wearing on our nerves, which the situation had already made raw. Meanwhile, our hotel was located next to the canal, at the center of the festivities—and the noise. Fantastic. That's all we needed.

Passing the time until my meeting with Lola, I meticulously prepared everything I would need for Mélodie and Bilel's last Skype conversation: two burqas, as he had demanded, my personal phone to record our conversation, the new one to contact him, and Mélodie's old phone, which I'd kept just in case. I'd lost a lot of weight during the past month. My face was emaciated. I'd gotten so wrapped up in this story. . . . From the bed of my shabby hotel room, I looked at Mélodie's djellaba and *hijab* for the last time. I felt a twinge of regret. Not because I was relinquishing my costume, but because I was about to abandon her, Mélodie. And I wondered, as I always do when coming out of periods of extreme adrenaline: *What now*? I wouldn't miss Bilel; that was for sure. Nor would I miss the shy and flirty way I'd had to behave with him. But daily contact with Syria would disappear from my life. I'd had to fact-check his claims, but Bilel had been my biggest source of information. Admittedly, I was addicted to this story. My thoughts were jumbled. I was tired.

To add to the stress, I received yet another call from my father. He'd been phoning all day. I didn't know what had gotten into him, but for the first time in my life I had a sense of what it was like to have a Jewish mother. That morning at the airport, we'd abruptly hung up on each other. He'd heard vaguely from my brother and mother that I was in Amsterdam, but that was all. He sensed I was avoiding him. It had barely been a week since I'd told him, without going into detail, that I was working on a special story. But he wasn't stupid. He knew my preferred fields of investigation. When he'd called, he asked what I was doing in Amsterdam and was especially eager to know where I was headed after Holland. He, who was usually so temperate and calm, yelled: "I heard talk of Turkey. If you go to Syria, you'll break my heart. Have you seen all the hostages in Syria? Have you? All journalists! Don't do that to me."

I was standing next to Charly during this parental explosion of worry. I didn't know if the famous war photographer could hear our conversation, but I was embarrassed to be fighting with my father "just" because I was going to the Middle East. Since my twenties, I've been careful not to expose my parents to the problems I encounter in my work. I also tend to avoid discussions about some of the trips I take while on assignment. My mom knows I'm careful, but my dad operates on the principle that danger comes from other people and it doesn't matter if you're careful. I spent several long minutes looking for a viable way to describe my situation, but floundered. I didn't see how I could tell him that his daughter had first become Mélodie, then Umm Saladîne, and was now married to a terrorist of the Islamic State, who also happened to be the French right-hand man of the most dangerous man in the world. "And Turkey?" he asked. He pretended not to understand that I was planning to go to the border. I was in over my head. Presently, in my hotel room, I hung up after promising my dad to be back in France on Tuesday, and reassuring him that for the moment I was simply in Holland. I would call back tomorrow so as not to worry him.

Everything was ready. While I waited for my partners in stress to return to the hotel, I checked my real Facebook profile, Anna's, to see if Guitone had replied to my last message. He had! That was one less thing to worry about. But then I read it.

"If you want an interview, enjoy some mint tea with the FSA."*

What had gotten into him? Why was he turning down my interview? What had I done? Only yesterday he'd been eager to meet and have his picture taken by my photographer. As a precaution against kidnapping, I'd told him we were preparing a big spread and had met several other Islamist branches in Libya (another jihadist nest). I'd brought my photographer, and we'd been treated really well. We'd been impressed by their ability to ensure our safety. The idea was that if Guitone and his brigade thought we'd met with other groups, they'd be less likely to set a trap for us. They wouldn't want to tarnish their image vis-à-vis other fanatics. That would be bad for them, since history tends to bring "evil forces" together. The Islamic State and al-Qaeda will undoubtedly end up joining forces in Syria.

I was overwhelmed. I was dizzy from not eating all day. Everybody who knew where I was had been sending me messages: my parents, Milan, my friends. I was red with embarrassment. I was hot. I opened the tiny windows, exposing the room and my head to the racket outside. My thoughts grew more and more confused, and my brain felt like it was going to explode with worst-case scenarios. I wasn't afraid of death, but I did fear rape. And according to numerous accounts, experiences of rape were common among ISIS's prisoners.

Guitone had been excited for my arrival in Kilis. He knew I was a legitimate reporter; he'd been able to verify my

* *The Free Syrian Army, the Bashar regime's primary opponent, and one of ISIS's enemies.*

identity online. We'd been corresponding for three months. Why, all of a sudden, was he backing out now? The "publicist" was usually so helpful. Why had he contemptuously suggested I speak with his adversary? What was going on? I hoped it didn't have anything to do with Mélodie. . . . If the Islamic State's French brigade made the link between Mélodie and Anna, I was doomed. Goodbye, Kilis. Hello uncertainty and doubt. I sent him a brief message letting him know that I was on my way, and that I'd be available for twenty-four hours once I reached Kilis. I reminded him that he'd given me his word—he'd sworn on Allah. Guitone never replied.

I waited for Charly to return before telling him the bad news. And also for him to give me the green light to contact Bilel. Our final conversation would take place in this cramped hotel room, whose features I was only just now noticing. The bed was gigantic, and there were no chairs. I would have to prop myself up against the pillows for our Skype conversation. The headboard was various shades of orange and looked like a kind of abstract rose. It added some Eastern flare to the scene. I hid any objects that might betray me, in case Bilel wanted Mélodie to show him the room. Charly and Hadrien returned, damp with sweat. They had found a perfect spot from which to take pictures and had spent two hours performing tests, when an orange-clad DJ showed up and mounted a gigantic stage there for himself . . . rotten luck. They didn't have time to find a better place. Our meeting time was approaching, and there was still no word from Lola. In the end, we came up with a simple solution. In keeping with our bad luck, we'd been assigned the smallest room in the hotel. However, it faced the road, and the window was well placed. Charly would be able to take the pictures from here. I'd simply make sure that Lola and I were positioned at a visible angle. That was a relief! Hadrien, who was very late for World Press, ran to take a shower. And finally, Charly gave me the green light I'd been impatiently awaiting.

Amsterdam, Friday, 6 p.m.

I put on Mélodie's clothes and sent Bilel my local number over Skype. Charly smoked a cigarette, enjoying himself as he watched me go about my rituals, such as worrying my favorite ring before removing it from my finger. He still called me Mélodie, even though he now knew my real name. He said the veil overwhelmed my face and wasn't very flattering. I smiled. His teasing lightened the mood, which I think is why he did it.

A video call from Bilel was coming in over Skype. My adrenaline shot up. Tomorrow we'd stand just yards from Syria. This call, which I thought would be the last, was the culmination of these past few weeks. My job was almost finished. My only fear in that moment was that Charly, seeing my conversation with the terrorist, would think I was insane. I hoped he would be able to distinguish between the journalist and the puppet. I sat up straight in bed and clicked the green icon. Bilel had so much to tell Mélodie; he didn't know where to start.

"*Salaam alaikum*, my darling, are you really in Amsterdam? I can't believe it. You'll be here soon. I'm the happiest man on earth. I love you, my wife."

I'd never seen him look so happy. His eyes shone with excitement. He was bursting with joy. Nothing betrayed his sincerity. Bilel was alone in an Internet café. He'd just finished "work."

"Yes, sweetheart. I'm here with Yasmine. We're flying to Istanbul tomorrow. But we have to be careful; it's not safe here. Tell me what to do."

As usual, Bilel was only half listening.

"You're so pretty!" he said. "Tell me about your trip. How did you pay for your tickets?"

"I stole my mom's debit card and bought two tickets online. We brought our passports, and here we are."

I forced a wide smile on my face to be more convincing. Mélodie had just left everything in order to be with him, and my attitude needed to reflect that.

"You're so strong, my wife! I'm so proud of you. You and your friend are both lionesses. If you still have the debit card, feel free to buy me some stuff."

"What do you want?"

"Oh, you know, darling. . . ."

I honestly had no idea. What do you get for a man who talks about how much he enjoys beheading people in one breath and how much he loves you in the next? A gun? Cash? Candy?

"Not really," I said.

"Well . . . cologne! But a good kind. You know, a good brand. I trust you to choose well."

Did he spray himself with cologne before killing people? That unnerved me. I knew that Afghanis perfumed the dead before burying them in white shrouds. Bilel had honed his skills in Afghanistan a few years before. . . .

"Which brands to you like?"

"I love Égoïste by Chanel, or something nice from Dior. I trust you . . . *Masha'Allah.*"

"Anything else?"

"Surprise me."

"Okay, baby. Can we talk about tomorrow? Yasmine is a little stressed-out, and she'd feel a lot better if she knew what was going to happen after we meet the mother."

"Oh, okay. Let me explain. When you arrive in Istanbul, you need to buy another phone. Throw away the one you got in Amsterdam. And be sure to pay in cash, not with your mom's card. Otherwise, the cops will be able to trace you."

"Okay. Where will the mother be waiting?"

"Actually, nobody will be there to meet you. You'll need to buy two tickets for a flight across the country; driving would take too long."

"What do you mean nobody will be there when we arrive? You promised!"

"I know, but it'll be okay. You're a big girl, aren't you, my wife? Dozens of Europeans make this trip every week. You can do this, my lioness."

"But that wasn't the plan, Bilel," I said, my voice genuinely frayed with anxiety. "We've gone over this many times. You were adamant—as was I—that a woman come to meet us. You told me we would be safe with the mother. How many times have you told me that nothing is more important than my safety?"

"Listen to me," he said, his tone hardening. "You're going to shut up for a minute and let me speak. It'll be a snap. When you arrive at the airport in Istanbul, buy two one-way tickets for Urfa. It doesn't cost anything, like fifty euros each. Pay in cash, okay? Otherwise, I'll pay for them; it's no problem. Before you leave Amsterdam, withdraw all the money you'll need and get rid of the debit card and the Dutch phone."

Urfa? Going there was suicide. Urfa and Kilis were almost equidistant from Syria. Only, Urfa was controlled by the Islamic State. Going there was the same as going to Syria.

Specifically, Urfa was the city where Guitone and his group ate kebabs, with Kalashnikovs slung over their shoulders and grenades tucked into their belts. Our plan was collapsing like a house of cards. I was beginning to panic. Meanwhile, I'd forgotten about Charly, who was slinking around the bed like a cat and taking pictures. I glanced up at him. He silently acknowledged that we were off to a bad start. I was suddenly worried, both about how this investigation would end and about Charly's freaked-out reaction to Bilel's words. My stomach was in knots. Improvising, I said the new situation made me uneasy. Yasmine was crying, I claimed, shifting blame onto the frightened minor.

"I'm okay, but Yasmine is flipping out. She's only fifteen. I don't want her fears to interfere with our plan. I already told you it's not safe here. I think we're being watched. I've made it this far on my own, but you promised you'd help me. So, I'm asking for your help."

His face contorted.

"Can you shut up with your crap for a second? Everything is going to be fine. Let me talk to Yasmine."

"No, she's scared right now. I'm her friend; I'll talk to her."

"I said, let me talk to her, Umm Saladîne!"

"I'll put her on. . . . Just give me some time. She's crying in front of the door to our room. Anyway, I'm not done talking to you. I think you're being unreasonably hard on me. All I ask is that you respect what you've been promising me for over a month. You say I can count on you. . . . Well, at the first sign of difficulty, you abandon me. That's great."

"You can't talk to me like that! Who do you think you are? I'm the one who gives orders around here, not you. Do you understand? Okay, well, show me your room."

Panic. The room was tiny. How was I supposed to show it to him without revealing Charly's presence? Charly, as discreet as a ghost, had been taking pictures with his Leica camera since the beginning of our conversation. We didn't have a chance to consult. It was a tense moment, and I couldn't even look at him. He crouched on his knees, contorting his body to stay out of sight as I slowly scanned the webcam over the room. Bilel was jumpy, and he carefully examined every inch of the hotel room. His anger seemed to be growing. He asked me to show him the room again.

"Where are you exactly?" he asked, his tone threatening and superior.

He'd never sounded like that before. Mélodie would have to apologize and try to calm him down.

"But, baby, I already told you," I said. "I'm in Amsterdam. You can see I'm not lying. I showed you our room. You saw my suitcase. Do you want me to show you the street outside?"

As usual, the terrorist wasn't listening.

"I want to talk to Yasmine. She has no reason to cry."

"Calm down, Bilel. Do you want to see my airplane ticket?"

I grabbed my ticket from my passport, praying he wouldn't ask to see it. Yasmine was already a problem I didn't know how to solve. That, plus Guitone, who had reneged on our interview, and I still hadn't heard from Lola. Nothing was going as planned. I'd dragged Charly and the magazine onto this sinking ship. That morning before boarding I'd shown Charly a few videos of our conversations. He'd seen that Bilel was a man in love, and therefore very dangerous. He'd been stunned to see Bilel's doe eyes for Mélodie. "Pathetic," he'd said. "And all for a religion whose name they're tarnishing." Bilel began to show his true colors. His tone became authoritative. His expression looked evil. For the first time, I heard other male voices around him. They also sounded threatening. I'd never seen him like this before. Suspicious. Unsettled. I didn't recognize him. He looked terrifying.

"Are you going to let me speak to Yasmine: yes or no? Do you think I'm an idiot? From now on, you're going to shut up. I'm part of a terrorist organization. You can't talk to me like that. Don't you know who I am? I command a hundred soldiers every day. I haven't even told you a quarter of the truth. I'm wanted internationally; that's why I can't even go to our cities in Turkey.* I can only travel to Iraq. I'm thirty-eight, you brat, and you and your friend can't bring me down. You don't know who I am. You'd better tread lightly."

He punctuated his words with a sadistic laugh. I'd stood

* *Cities occupied by the Islamic State.*

up to him. Giving Mélodie a little character had been a miscalculation. Bilel didn't appreciate it at all.

"I would never try to trick you," she said, submissively. "It saddens me that you think I'm running away to another country in order to hurt you. I don't know what to say. I want to cry. Whatever you want, I'll do it. I'll get tickets for Urfa. I'll do whatever you say. I promise."

"I'm really disappointed in you. I thought you were a strong woman. Let me speak with Yasmine; I'm not going to bite her head off."

Night had fallen, and I didn't know how I'd find a teenage girl capable and willing to play a would-be jihadist. Charly might have been handsome, but he couldn't pass for a woman—not even if he wore a wig. My only option was to continue apologizing in an attempt to calm Bilel. Slowly, Mélodie did regain the killer's trust.

"If she's incapable of taking two planes, you don't have to bring her," he said, sighing. "Shit, ditch her. She can go home. I don't care about her. Do it! Get rid of her."

"I'm not going to abandon a sister. Don't be angry. I understand your instructions. I'll reassure her, and tomorrow we'll go to Urfa."

"There's the wife I recognize. Okay, I'll call you back in ten minutes with instructions on what to do once you arrive in Urfa. Believe me, you'll be well taken care of."

Friday, 9 p.m.

I tore off the *hijab* and rose to my feet. I paced the tiny room, my face in my hands. Everything was falling apart. I hoped I'd regained control of Bilel, otherwise our story wouldn't hold together. Lola had stood me up, Guitone was setting me up, and Bilel had asked me to go to one of the most dangerous cities in the world. Charly kept quiet, waiting for me to break the tense silence.

"This is bad, isn't it?" I asked, turning to face him.

Charly nodded.

"Isn't this making you schizophrenic?" he asked, troubled. "I've seen some things in my life, but *that* . . . I'm impressed by your self-possession. Bilel is insane. One minute he's talking to you about cologne and showering you with affection, the next he's threatening you and ordering you to ditch your friend."

For Charly, the stakes of this story had just become clear. I knew he'd seen and heard much worse, but he really looked shaken up. The passing minutes seemed endless as we waited for the terrorist to call back. We reviewed potential scenarios to fill the time. First of all: would he call back? It would be disappointing if he didn't. We were both freelance, and we wanted to deliver what we'd promised. All our plans were falling like dominoes. To myself, I thought, here is the omen's destiny.

Twenty minutes later, Skype rang. Charly and I had almost given up hope. We held our breath, unsure whether the call would be fruitful or disastrous. I was smoking a cigarette by the window and speaking with my editor in chief, who had orders to wrap up this story over the weekend. When Bilel's call came in, I was relating recent events. I immediately hung up and hurried to the bed, almost slamming into Charly. I answered. Bilel appeared on the screen. He looked calmer. He was even smiling, his eyes widening when he saw me. In my rush, I'd forgotten to put on my veil. How stupid of me! Luckily, the room was dark and my hair was tied back in a ponytail.

"Did you take off your veil?" Bilel asked, visibly delighted.

"Yes, just for a second. I went downstairs to get a Fanta. Remember: I don't want to draw attention to myself. You called just as I was coming back to the room. I would have missed. . . . Give me a second; I'll put it back on—"

"Don't do that!" he interrupted, his eyes shining. "It doesn't bother me if it's just you and Yasmine. *Masha'Allah*, our children are going to be really good-looking with parents like us. *Insha'Allah*."

Bilel was so modest. I smiled at him and shot Charly an exasperated look. My colleague rolled his eyes.

"Yasmine is feeling better. She's downstairs now, but I convinced her to take the second flight tomorrow. So, how does this work?"

He didn't answer right away. Instead, he pushed his chair back to get a better look at his future wife, feasting his eyes on my face and biting his lip.

"We should talk about our wedding night, too."

"Let's wait until it's just the two of us. . . . It's too personal."

"Okay . . . I hope you have some pretty underthings for that special night. Remember what I told you: anything goes with your husband."

"We'll see. I don't feel comfortable talking about that, Bilel."

"I understand. Anyway, we'll be together tomorrow night. Less than twenty-four hours separate us, my love."

"It seems so far away. What do I have to do in the meantime?"

"There's a problem: the phone networks where I'm located are bad. So, first you're going to use your new phone to call the number I'm about to give you. Tell the person who answers that you're Abu Bilel al-Firanzi's wife, and that you're calling on behalf of Abu Omar Tounsi in Syria. Then, give him your flight number and arrival time in Urfa."

"Okay. Who will I be speaking with?"

"Don't worry about that. Here's the number. I want you to call now, with me here."

He recited eight numbers for a Syrian phone number and watched as I dialed. A man answered in French and asked who was calling. Mélodie repeated exactly what Bilel had told her to say. The man on the other end of the line told me to take a domestic flight from Istanbul to Urfa, since "the roads are being monitored, especially with a minor."

Mélodie agreed. He asked if she needed money to buy the tickets. She said she could pay for them herself. He encouraged her to call him "at any time day or night until you arrive at the border." Mélodie thanked him and hung up. She turned to look at Bilel on the computer screen.

"Are you sure you hung up?" he asked.

"Yes."

"Okay. Good job. You followed my instructions. *Masha'Allah.* You'll have to call him with your flight details tomorrow. Now I want you to call someone else. It's the guy in charge of border safety. Tell him you're my wife and you'll be treated like a queen."

"You won't be there?"

"No. I already told you, I can't go to Turkey. But I'll be just a few yards away. Don't worry, my wife. Once we're together, I'll be at your side forever."

Awesome! I thought. Meanwhile, I realized that even though I'd bought the maximum number of prepaid minutes for foreign calls on my Dutch phone, I was already running out. I still had Mélodie's phone, but it was also almost empty. A phone with a French SIM card calling Syria from the Netherlands wouldn't last long.

"Okay. Tell me what I have to do."

Bilel gave me another Syrian number. No answer. It didn't matter. Bilel had a solution. He knew somebody else I could call who would definitely pick up. The only problem was he didn't speak French—only Arabic. Bilel asked if Mélodie could manage. She said she was only able to hold a simple conversation, and this would be too complicated. Again, no big deal. Bilel instructed Mélodie to dial the number and switch on speakerphone. Bilel would speak with the man himself. Just an hour before, he'd screamed at Mélodie to "shut up." Now he was offering precious information that could be used against his branch.

Bilel had brainwashed Mélodie, renamed her, and was now using her as a proxy from her hotel room in Amsterdam. Although what I experienced could never be compared to what a person faces in war, it was still dizzying. I didn't have a choice. Mélodie dialed, but there weren't enough minutes on the phone and the call didn't go through. It was late. The stores were closed. No matter what I did, I wouldn't be able to find a safe phone. I still had my phone, which I'd been using to record our conversations. But that would put me at further risk. Although my number was unlisted, I knew that anyone with the right contacts could trace it to me. And the Islamic State has very good contacts. In normal circumstances, I would have ended things there. I'm not reckless. I tend to play it safe. It's just . . . I was so close. All I needed was a phone. Which I had in my hand . . . Oh well. When I returned to France, I'd change my phone number, which I'd had since I was sixteen. I'd sign up with a competing operator. There had been too much failure today.

I dialed the number, putting the two men in touch. Their conversation lasted about three minutes. Although Bilel still had an eye on Mélodie, I was able to exchange a look with Charly. Without speaking, he asked if I was okay. I signaled yes. Later, a girlfriend at the magazine who speaks fluent Arabic translated the conversation. Among other things, Bilel adamantly conveyed that I was in possession of a French passport. After their discussion, Bilel devoted all his attention to his wife. That was when Hadrien knocked. The poor guy had been running around since early that morning. It was now almost ten o'clock, and he'd hoped to get a half hour's rest. Charly silently opened the door and signaled not to make a sound. Yet another person to see me playing Mélodie. Hadrien had looked at pictures of me wearing my costume, but seeing and hearing me live was another experience, as I'd discovered with Lou in Tunisia. Over the course of my conversation with Bilel, I'd put on my veil. Hadrien was like a chameleon. He avoided looking at me, leaned against the wall, and lit a cigarette. It was just as well that he didn't pay much attention to Mélodie. He was unaware of everything that had happened; particularly Bilel's terrifying show of anger. Later, he'd tell me he hadn't been able to look at me; it was too disturbing. He was suddenly faced with the culmination of his fears for me in that moment when I was someone else.

The fighter had gotten what he wanted, and he was relaxed. He laid on the sweet talk, saying he couldn't wait until tomorrow. He was especially adamant his wife call him as soon as she arrived in Istanbul. Even if he couldn't be there himself, he'd make sure she was safe. I couldn't bear him anymore; all I wanted was to hang up. The three of us were suffocating in that tiny room, and there wasn't any bottled water. We couldn't even open the window because of the noise outside. Mélodie tried to end the conversation, but Bilel ignored her.

"Don't forget: I need cotton boxers. The ones here are too rough."

I had completely forgotten. This was exactly the type of useless request that irritated me. I couldn't listen to him anymore.

"Right, right. Of course."

"Size large, okay, honey?"

"Got it."

"Okay. And do you have the scarves for the *qadi* who's going to marry us?"

"Yeah."

"Okay, well, max out your mom's debit card, and don't forget to take out some cash. And bring me presents! I can't believe you'll finally be here tomorrow!"

"Yeah . . ."

"Be discreet, okay? A lioness!"

"Okay."

"Are you happy?"

"Of course!"

"Are you scared?"

"You told me not to be, so I'm not."

"But what about for our wedding night, are you scared?"

"We can talk about that tomorrow."

"Yes, we can. Leave everything behind, except some clothes and intimates. We can get everything else here. Everything except underwear. Don't forget to bring some nice lingerie."

"Okay. What about my computer? I stole my mom's."

"Even your computer!"

Then he reconsidered.

"What type of computer is it?"

"A Mac. It's pretty new."

"Go ahead and bring it. Just be sure to erase your old life before tomorrow morning. And don't crack! Don't send anything to anyone. Nothing can keep us apart now, Umm Saladîne. So keep an eye on Yasmine."

"Okay. See you tomorrow, Bilel."

"I'm the happiest man on earth. You're mine now."

Mélodie smiled. It was probably her least convincing smile since the beginning. I was exhausted. I was sick of this crazy man, of this month, and of this day. Finally, she hung up. I let out the longest sigh of my life. I felt like I was walking on a high wire. I knew I hadn't become Mélodie, but I had put myself in the skin of a tightrope walker who was afraid of heights.

Friday, 10 p.m.

Charly and I were both shaken up. Hadrien suggested we eat out. I had only a few minutes to regain my senses—and my identity. I changed clothes and let my hair down. Hadrien found us a charming restaurant that looked out onto the canal. We ate quickly, and I finally felt like I could breathe.

My friend sat across from me, and next to me was Charly, whom I'd just met but who already seemed like a friend. I slowly regained my self-confidence, but I had no appetite. Bilel hadn't discovered my true identity, but everything had fallen through. We still didn't know if we were going to Kilis the following day. Deep down, I already knew we wouldn't. On our way to the restaurant, I'd described the situation to my editor in chief. I bitterly laid out the facts: we had the choice between flying to Urfa, which was out of the question, and going to Kilis to interview Guitone, who didn't seem interested in talking to me. It was basically a choice between being kidnapped in Urfa or running a 30 percent risk of being kidnapped in Kilis. My editor gently replied that we weren't going to take such a big risk just so I could describe life in Kilis. Besides, we already had plenty of information for my story. We would speak again in the morning.

To put things in perspective, she reminded me that Édouard Elias and Didier François, two journalists sent to the region by Europe 1, a French radio station, had just been freed after ten months of captivity at the hands of ISIS. Not only did I feel as if I'd failed; I also felt remorse for Mélodie. I owed her a dignified ending. My colleagues tried to distract me. Our dinner conversation became more personal. All three of us needed to let off steam, and Charly shared some of his experiences. Some were hilarious, others sad. I didn't say a lot, but I drank some wine. The only problem is, I can't really drink. After a few glasses I become an embarrassment to my friends. Disinhibited, I talk loudly and make a scene. I'm basically a really annoying drunk. That's why I rarely drink.

That night, I forgot my limits. I couldn't get the thought of the morning's uncertain trip out of my head. Time was running out. It was Friday. I had to finish the article on Monday. I had a lot of work to do that weekend to fit a month's worth of experiences into one piece. I was stressed-out. I drank. I lost control. Hadrien noticed something was wrong. He insisted I come with him, if only for an hour, to World Press's gala. I'd already decided I wouldn't go, but he thought it would help clear my mind before I went to bed. I don't typically like mixing work and pleasure. I'd barely eaten, and my head was spinning. The Bordeaux ended up deciding for me. With the wave of a magic wand, I found myself in the middle of two hundred partygoers. I ran into some people I knew, and I began to relax as I drank more and danced. Things were looking blurry, but I had the presence of mind to check my watch. 2 a.m. It was time to go back to the hotel. I told my team I was leaving. "We'll meet you in the coatroom," they said.

I don't remember much after that. I don't know what I did to annoy the bouncer, but I do have vague recollections of being restrained by the man. He'd hit me and I'd tried to fight back. He was two times my size in height and girth. Everybody was yelling: Hadrien, Charly, my friends. The bouncer slapped me, after which all I remember is anger coursing through me. I aimed my foot at his crotch. Hadrien restrained me with all his strength, trying to prevent me from doing more damage. I think he whispered: "Stop it. You're stressed-out. Come back with us. Everything is going to be okay." Then a large group of people appeared, including the event organizer. Apologies were exchanged. I looked for Hadrien but couldn't find him. Charly accompanied me back to the hotel in a taxi. The drive seemed endless. I still didn't know where Hadrien was. I don't remember if I asked Charly.

Saturday morning

I only remember the awful feeling of waking up. Too much alcohol and too many murky memories. Flashes of the evening slowly came back to me. Then I recalled the discussions with Bilel. In addition to my headache, I had a sore arm, and my shoulder was throbbing. It was nine o'clock. I called Charly. He said he'd just taken Hadrien to the airport. I took a shower and met him outside the hotel. We walked for a while, with strained faces, until we found a café a bit calmer than the others, which were still playing the same deafening music from the previous night. Before discussing the evening, we called our editor. She'd spoken with the magazine's managing director and assistant director. They all agreed: we should return to France. It was another blow. She explained that we already had enough elements for an amazing story. The rest would have been a bonus, but we didn't need it.

I knew she was happy with our work, but I also knew she was trying to console us; she'd worked in the field for a long time and was familiar with what we were feeling. I sent a brief text message to my parents telling them not to worry; I'd be in Paris that evening. I felt bad, especially for Charly. He'd been excited to work on this story, and he hadn't gotten to contribute much. I told him it was my fault. A series of unfortunate circumstances had piled up: Guitone, Lola, Bilel's lies. It didn't make a difference. He was still disappointed in himself. And I was disappointed in myself. Internally, I cursed Hadrien for abandoning me the night before. I was surprised: he was usually such a loyal friend. Timidly, I asked Charly how Hadrien was doing. Apparently he was sore, and worried about us. Sore? Why?

"You probably don't remember, but when you tried to hit that angry bouncer, Hadrien threw himself between the two of you—to protect you. He's the one who got hit."

Now I really felt miserable. I tried to act calm, but I was ashamed and could feel tears welling up in my eyes. I wished I could disappear. Be as small as a mouse—anonymous.

We canceled our tickets for the nth time. On the phone, the girl who manages our travel was tearing out her hair. I'd changed our reservations with each development, and now I politely asked her to cancel everything and book us two tickets on the next flight to Orly*. Forty-five minutes of travel instead of the five hours we'd planned for the day. On our way to the airport, we received text messages of congratulations and support from our superiors. We had nothing to report, and we both felt like frauds. I also received a very touching message from Hadrien when he landed in Paris. He didn't know that we were on our way back to France. He made no

* *Translator's note: Orly Airport, near Paris.*

mention of the previous evening. He simply wrote that when he saw me wearing Mélodie's clothes, he realized I might go to the border, and he didn't want to part on a bad note. Our friendship was too precious. And life even more so.

As we waited to board, Charly watched videos of Francophone fighters for the Islamic State on my computer. He knew what was going on in Syria and elsewhere. But this was the first time he'd seen the Internet propaganda that had inspired my story. Stunned, he wavered between hysterical laughter and expressions of dismay. I remembered reacting the same way when I first saw these videos. While he watched, I went to smoke; I knew these videos by heart. I called my father from the smoking room. Now that we were headed home, I owed him an explanation. I briefly summarized the events of the past month. He interrupted me at every other word. "You're crazy, Anna!" he exclaimed again and again, his voice frayed with worry. When I was done, I told him it was all over. He replied that soon Prince William and Duchess Kate would be baptizing baby George. He wondered if that could be my next story. He was joking, and I was glad to hear him laugh.

I found Charly, and we boarded. He started calling me Anna. It symbolized the end of our investigation. We spoke again of some of the confidences shared the previous night over dinner, then Charly dozed off. I rested my head against the window and gazed at the clouds. No music this time. I wasn't in the mood. I considered the writing that awaited me and the mass of information to be condensed into ten pages in just twenty-four hours. Then my thoughts wandered to the night before. I was surprised by my behavior. I'd never gotten into a fight before. For some reason, the bouncer had unleashed all my bottled-up emotions about Bilel, bad omens, failure, my digital double. I don't know, but in that moment, I was forced to admit that I really had become schizophrenic.

Paris, Sunday afternoon

It was about 3 p.m. I raced against the clock to finish my article. Some aspects of my investigation were extremely complex, and I struggled to condense everything into the allotted page limit. It wouldn't be a problem if my deadline were still tomorrow. But Monday was a holiday, and the deadline had been pushed up to today. Without rereading what I'd written, I sent my story to the features editor. We had a "hot and cold" relationship. Sometimes he leaped to my defense, fighting for one of my stories. At other times he replied to my work with a terse email, followed by harsh criticism in a staff meeting. I often loathed him on the day a story was due, but he was really experienced, and I owed a lot to him. Now that I'd submitted my article, I considered going to bed and hiding under my covers like a little girl who's brought home a bad report card. I always feel that way when I finish a story. What a brave reporter I am. But this time, the features editor didn't reply via email; he called. That was usually a good sign. It meant he wasn't unhappy, and he'd congratulate me on a job well done, in sober and perfectly chosen words. Now I sensed he was more excited than usual. He encouraged me to keep writing. He didn't care if it was too long. This story was worth developing. "The devil is in the details," he said. That was true. He was as fascinated by Bilel's character as he was by Mélodie's courage. His only advice was to emphasize how intensely I'd entered into my character. He said I shouldn't be afraid to put myself in the story, so long as I maintained a neutral and professional tone.

I'd been writing nonstop all weekend. He knew that, and he sensed I was losing perspective. He encouraged me like a good teacher: it would be a tight deadline, but everything was there; I simply had to write it down. I regained confidence in myself and continued writing my first-person narrative. I received text messages from my editor in chief, the assistant director, and even the magazine's lawyer. The features editor must have updated them on my situation. Had they already read what I'd sent him? According to them, I had "a bomb on my hands," and we needed to talk about it prior to publication. I wasn't sure how to interpret their messages, so I left for the magazine's offices, lugging my computer and recordings. I lobbied for my story, going from office to office and pitching my unfinished article to every editor who might be interested. I'd never done something like that before. I even knocked on the director's door. I'd practiced a short speech on the metro: "I know you're busy and we're coming up on a deadline, but please just read the introduction to my story. It'll only take ten minutes." But when he saw me, he guessed the reason for my visit and didn't give me a chance to speak.

"Are you here to talk about your story?" he asked, sounding busy.

I timidly said yes and handed him a few printed pages. Then I went to my office, crestfallen. The editor in chief, who had been following this story since its inception, came to see me. She'd read the article. She thought it was good. She was sorry, but she hadn't realized there would be so many elements to review, particularly from a legal perspective. Although she'd been aware of my activities, she hadn't understood until now how much I'd invested myself in this story. Then the assistant director came into my office, closing the door behind him. He was familiar with all the issues raised in my story, even more than I was. He was impressed with my work—that wasn't the problem. He hadn't realized that Bilel was such a big fish, nor had he been aware of the terrorist's close ties to ISIS's leader, al-Baghdadi. There was more to consider than the threats of a man who felt used. Bilel had divulged information on location and strategy. The repercussions could be serious, and we had to be extremely careful. There were too many unresolved issues, and we were just hours from going to press. We needed time to think, so we delayed the story's publication by a week.

Back at home, I had mixed feelings. I was relieved to have extra time to polish my writing, but I was frustrated not to have "wrapped up" the story. I also wondered what had happened to Bilel. I'd been so busy writing about him and fighting for this story that I'd forgotten about him. I hadn't checked Mélodie's accounts for twenty-four hours, and her various disposable phones had been off since our departure from Amsterdam. Mélodie had sent Bilel a Skype message from the airport informing him that a "strange" man had questioned the girls. Yasmine and Mélodie felt they were being watched, and they decided to return to France until better circumstances presented themselves. Besides, she admitted, Bilel had been right, Yasmine was a problem. Her family wouldn't stop calling. Mélodie would make the trip alone, but for now she didn't want to endanger her man or his brigade. She would lie low for a while in Toulouse. Given the situation, that was the best solution for everyone.

I plugged in all my devices to see how Bilel had reacted. The Dutch phone had been bombarded with messages. Most were from men I didn't know, except Abu Omar Tounsi, the man "in charge of security at the border." Everyone was asking where Umm Saladîne was . . . including her husband. The tone of his messages reminded me of our heated conversations in Amsterdam. One line in particular stood out: "Where are you, you little bitch? I swear to Allah, you're going to pay!" I turned off the phones and decided this was the end. It was time. I'd suppressed my own character to make Mélodie submissive. Bilel wanted to threaten Mélodie? Anna would reply from here on out. I checked everything one last time before making her disappear. Bilel had sent his usual litany of messages on Skype:*

"Where are you?"

"Where are you??"

* *One can receive written messages on Skype, even when not connected.*

"Where are you???"

"Where are you????"

"Where are you?????"

"Where are you??????"

"Where are you???????"

"Hey! Where are you, bitch??????????????"

His anger could be measured by the number of question marks he used. Mélodie received a single message on Facebook, which at least got to the point for once:

> *SHIT, WHERE ARE YOU? I told you to deactivate this account. You're on your own now. I'm really disappointed in you. You're not a lioness after all.*

It was just as well. I preferred knowing he was angry with Mélodie. If he got sick of her, he'd forget her. I hastened to deactivate my avatar's virtual existence, only keeping her Skype profile. Mélodie used it to send a final message, so that her sudden disappearance wouldn't arouse suspicion. She apologized, but her return to France had been rough. Her mother had called the police while she was in Amsterdam, and she'd had trouble justifying her absence. They'd confiscated her computer and telephone, and, as Bilel had instructed, she'd gotten rid of the new phone bought in Amsterdam. Besides, it was better if they cut off all ties for the moment. The lid could blow off on anyone at any time. She apologized again, and said she was sorry she wouldn't be able to communicate with him anymore. These were her last words to him:

I'm sorry, Bilel. I never wanted to disappoint you, but when I sensed danger—for us and for you—I thought turning around was the best course of action. I hope you'll still want to talk to me once I have a new and secure phone or computer. With love, Mélodie.

I had no intention of getting back in touch with that dangerous lunatic. But I hoped to convince him of Mélodie's good intentions and curb his anger. The more Mélodie showed remorse, the more she regretted how things had turned out, the easier it would be for Bilel to move on. After all, he had more important things to do than think about Mélodie, who was just another twenty-year-old girl. The Islamic State was preparing its assault on Iraq. In almost two months to the day, ISIS would seize Mosul, the country's second-biggest city, before turning its attention to Baghdad, which would awaken the world to the diabolical specter of the organization's radical fundamentalism. It was an enormous project, and he wouldn't have time to think about Mélodie.

Or so I thought.

Two days later, at the magazine

There was less pressure, but I still felt tense. Mélodie had vanished, and I had no idea if Bilel was still angry. Anna, the journalist, had an article to finish. Yesterday I took stock of everything that had happened over the past month. I also let myself rest. Today my head was clear and I was anxious to end this story. The last lap is usually the hardest. After finishing, I'd trust the magazine's judgment on how to proceed. I was writing in an office, surrounded by my friends and colleagues, when I got a call from a French cell phone number. I answered. It was Bilel. I jumped from my chair and ran down a hallway. How could he be calling me from a French number? And on my personal phone? I remembered being forced to use my own phone in Amsterdam. One misstep and my cover was potentially blown. At least that's what I thought. I had trouble understanding what the terrorist was saying. The connection was bad, like when Mélodie used speakerphone to mediate Bilel's conversation with his Arabic-speaking contact. I knew panicking wouldn't help. Bilel couldn't have returned to France over the past forty-eight hours. So, although I'd just retired her, I pretended to be Mélodie again. Bilel asked me where I was and "what the hell" I'd done. Mélodie repeated more or less what she'd said in her "goodbye" message, adding one important detail: her mother had found their messages and sent them to the cops. So he shouldn't call this number, not even in an emergency. She'd disconnect it soon.

"Are you threatening me now?" the terrorist asked, his tone condescending. "That's too much!"

"Of course not," Mélodie replied. "I'm trying to protect you—"

The conversation ended abruptly. I found myself hiding in a co-worker's office, holding my phone, wide-eyed. I immediately went to speak with the editor who'd been working closely with me since the beginning. In addition to being the person who oversaw my work, he'd also been a lifeline and important source of advice in this atypical investigation. He jotted down the number from which Bilel had called and asked me to wait a minute while he traced it. I returned to my computer and signed in to Mélodie's Skype account. I wanted to scramble the IP address and see if Bilel had left any threatening messages. There were several.

"Who do you think you are, you little bitch?"

"You've underestimated who you're dealing with . . . a terrorist organization!"

"The people you talked to last weekend have fifteen years of counterespionage experience. We'll find you in a matter of minutes."

"You wanted to make an idiot out of me; I'll make you pay. ROFL."

This time I felt as if a blade were at my throat. Although

the threat was distant, it was becoming more concrete. I disconnected without replying. As promised, my boss found me and pulled me aside. The French number belonged to someone called Hamza,* who lived in Albertville, a town in the Savoy region. We gazed blankly at each other, tensely searching for a logical explanation, before bringing the news to the editor in chief. The three of us discussed the issue in her office. She decided to call the number pretending to be Mélodie's mother. Strangely, there was no answer. We found a landline corresponding to the cell phone's owner in the white pages. An elderly man answered the phone and said he was Hamza's father. My editor told him she didn't think it was appropriate for his son to call her daughter, who was only twenty years old. The man didn't react. But as soon as she mentioned Syria, he panicked. "My son is an adult," he stammered. "He makes his own decisions!" Besides, he hadn't seen his son in weeks, and he didn't know where he was. Mélodie's mother said she was surprised he wasn't more worried. He quickly ended the conversation, as if suddenly seized with panic.

The article would soon be written, but for me the story was only just beginning.

---

* *The name has been changed.*

The same day, in the evening

Back at home, I stared at my black sofa. I didn't recognize it anymore. I hated it. My phone rang, drawing me out of my stupor. Another French number I didn't recognize. I asked who was calling. A young-sounding male voice politely replied that this was Hamza's brother. *Now what?* I wanted to scream. I didn't feel like resuscitating Mélodie, and I couldn't tell him I was a journalist. He sounded younger than me, so I made myself older.

"Well, this is Mélodie's mother!"

Silence. I continued:

"What does your brother want with my daughter?"

"I don't know, ma'am; I promise. My brother disappeared a few weeks ago, and I haven't heard from him."

"And you think I have? Forget this number and forget my daughter!"

"I don't understand why he hasn't contacted anyone, but your daughter . . . if you could tell me more about her."

The brother sounded truly distraught. His voice quavered. His words were jumbled. He probably had suspicions, but he didn't really know where Hamza was. But Mélodie had already given enough of herself, and her mother wasn't going to take her place. My voice, no longer that of a mother hen, now took on the traits of an Italian mamma with a whole family to back her up.

"You listen to me: first I got a call from your little brother, then your father, and now you. *Khalas!* I've had enough of the Hamza family. I'm warning you: if you want your family line to continue, you'd better figure out a way to have him call me by tomorrow morning. Otherwise, I'll make sure Domestic Security pays you a visit in Albertville. I'll also send my brothers. And believe me, kid, I have a lot of brothers."

I smiled to myself as I hung up the phone. It felt good not to be the submissive and frightened Mélodie anymore. The next morning, some people at the magazine informed me that the Hamza family was now unlisted. The authorities were indeed looking for Hamza. He'd left France three weeks before; his last known location was in Turkey. Since then, he'd become a ghost. None of this was reassuring. Either Hamza was in Syria, and I had nothing to fear for the time being, or he was in France, and maybe even in Paris. We had no idea. A stream of numbers beginning with +591, +886, and +216, all Syrian and Turkish prefixes, kept calling. I didn't answer.

Less than twenty-four hours later, most of the dailies and news channels headlined this: "six people living in Albertville between the ages of twenty and thirty-seven have been jailed for their involvement in jihadist recruitment networks." Tomorrow would be May 1, and I wished with all my might that this was all a very bad—and late—April Fool's joke. I didn't know then that the police were tapping "my" phones. I discovered that three weeks later, when my name appeared in several legal cases related to departures to the Middle East. Among those targeted was Vanessa, the girl who was six months pregnant. After our exchange of text messages, Bilel had told Mélodie not to count on Vanessa because she'd "chickened out." Indeed, the young girl who'd been so impatient to give birth to her child in Syria could no longer be reached by phone or even on Skype. Our exchanges helped the authorities put together a solid file on Vanessa, and they stopped her before her departure. At the same time, they dismantled an important recruitment network linked to the girl.

Without knowing it, Mélodie was accumulating enemies. When we learned about the arrests in Albertville, everybody at the magazine—and especially me—wondered if my story had played a role. I'd encountered so many coincidences and competing circumstances since the beginning of this story, but as a reporter I didn't believe in chance. The magazine, citing Bilel's threats, the Hamza family, and now this, asked me to change my address and telephone number as soon as possible. I had to leave. Now. Right away. If ISIS made the connection between Mélodie and the journalist, and if it blamed me for the recent spate of arrests, my life would never be the same. I couldn't believe it. Even if the DCPJ had gleaned valuable information by listening to my phone calls, I had trouble believing I could be the one responsible for sinking the Islamic State's recruitment networks. Still, I had to be careful. My first instinct was to throw a few essentials into a bag and seek refuge with my parents. I figured I'd spend one or two nights in my old bedroom. Six months later, I still find myself staying there from time to time.

Five days later

I was still living there when the ax fell. That morning, after careful assessment by the legal department, the magazine sent my article to press. Meanwhile, Charly and I rushed to the Nigerian embassy for visas. The recent kidnapping of more than two hundred girls in Chibok, a small town in eastern Nigeria, by the Islamist group Boko Haram had elicited an international outcry. We had to get there, but Charly and I quickly realized that we wouldn't be leaving France. There was a one-month waiting period for journalist visas. Besides, the director of the magazine sending us to Nigeria called to tell me that my going to that kind of region was out of the question. I swallowed my disappointment, telling myself there would be other opportunities to go into the field. It was presently 7 p.m. The weather was still pleasant and warm. Lou had come over to spend the afternoon with me. We were lazing in the sun. I felt light. Everything was fine. I still received phone calls from strange numbers, but nothing more, and I hadn't reconnected to Mélodie's online world. Everything was deactivated. Ignorance was bliss. The article would appear in two days. If things remained calm after its publication, I could return home and get back to my normal social and professional life. Like me, Lou was taking a break from her hectic schedule. We were laughing about something or other when the magazine's editor in chief called. I calmly answered, my voice relaxed. She repeated my name a few times, as if to be sure that it was really me on the other end of the line and that I was listening.

"Anna, Bilel is dead."

Silence.

"Bilel is dead!" she said again. "Do you understand?!"

No. Not really. My poor editor had just unwittingly torn me from the cocoon in which I'd been living for the past several days. My head spun and my body rocked. I got to my feet and walked without direction, trying to concentrate on what she was saying. Apparently, David Thomson, a very reliable journalist for RFI* and a specialist on religious fundamentalism, had announced Abu Bilel al-Firanzi's death over Twitter. The tweet included a picture of Bilel from when he was still alive. I'd actually seen the same photo before. David Thomson rarely made mistakes. My editor, happy for me, was excited to give me the news. In her shoes, I probably would have been, too. The death of a man, even if he was a murderer, didn't excite or please her. She was thinking of me, and she assumed that Bilel's death would reduce, if not eliminate, any risk of retaliation against me. She cared about me and she was relieved. That was why she got a bit carried

* *Translator's note: Radio France Internationale is a French news radio service that broadcasts all over the world.*

away. I didn't say anything. She sensed my malaise and quietly asked if I was okay. I said yes, just a little shaken. It would pass. I told her I wanted to see what Mélodie's friends had to say. I would keep her updated. As I went back into the apartment, I wondered when the omens and their destinies would finally reach their conclusion.

Lou understood what I was feeling. She followed me into the kitchen, where I wanted to check the Internet. She advised against it. I was edgy. My hand trembled like a leaf. She hugged me, and for the first time since this had all started, I broke down. Lou knew I didn't have any feelings for Bilel, but she asked if his death upset me. I was crying like a little girl, tears streaming down my cheeks. I didn't care how Bilel had died—that was his problem—but I did want to know why. If his sudden death had anything to do with me, I would hate the role I'd played. He was a dangerous murderer, but I wasn't. The idea of having any connection to the death of any man, no matter who he was, horrified me.

In that moment, I felt as if *I* had sentenced him to death. I'd first become Mélodie, then Umm Saladîne, then Mélodie's mother, and now an executioner. I'd never intended to play that role. Maybe the terrorist organization, discovering how I'd deceived and gleaned information from Bilel, had punished him. And although Bilel embodied everything I loathed, the announcement of his death was deeply disturbing. Meanwhile, the news had begun to circulate, and I started receiving text messages practically congratulating me on the sudden death of my "husband." Nobody who wrote took death lightly. They were all simply interested in my well-being. They were trying to be thoughtful. But I didn't know how to respond. And I was baffled by the flood of tears flowing from my eyes and the gut-wrenching ball in the pit of my stomach.

I calmed myself down and connected to my real Facebook account, where I had some contacts with ISIS fighters. I needed to know how Bilel died; I was obsessed. The context of his death would tell me if I'd played a role or not. I wouldn't feel anything if it wasn't my fault. Or rather, like my friends, I'd be relieved. I spent an hour or two scanning the pages of various ISIS mujahideen. Many paid their respects to "Abu Bilel al-Firanzi, who served in the name of Allah for fifteen years. He was the closest Frenchman to our caliph, Abu Bakr al-Baghdadi." I noted in particular Abu Shaheed, a typical Frenchman working for ISIS in Syria, who has been interviewed on numerous occasions by journalists over Skype. His cover photo was taken somewhere in Syria and showed him pointing a gun at the camera. The caption read: "I'm aiming at Domestic Security." Abu Shaheed is an extremely influential leader among French jihadists. His reputation more than preceded him. A few months before, he'd closed his Facebook account, saying that faith wasn't compatible with superficial social networks. Today he'd decided to reactivate it in order to announce his good friend's death.

"I'm only reconnecting to pay tribute to a great brother. It is with deep regret that I announce the death of our dear and esteemed Abu Bilel al-Firanzi, who devoted his life to Allah. He went to negotiate a peace deal with al-Nusra near Iraq, but those *kafirs* set a trap for him. He was the closest Frenchman to our emir Abu Bakr al-Baghdadi. Now he is happy because he is with Allah, as he always wanted. Let's be proud of him. Please send your thoughts to his wife, and especially to his children."

Bilel had been "negotiating peace"? Him, a man of peace? He and Nelson Mandela. They should put that on his tombstone. I was becoming more and more hysterical with each passing minute. I found a video of an explosion on YouTube and other specialized sites; apparently this was how Abu Bilel met his death. The clip showed the earth literally lifting into the air. It was impressive. According to the video, the al-Nusra Front had planted a bomb in a tunnel between Syria and Iraq. Bilel was on a secret trip "to sign peace treaties . . . but they ambushed him." The bomb also killed a considerable number of civilians.

Two French voices commented over the video, which lasted less than a minute, congratulating themselves on a successfully executed trap. It was credible. Bilel had often mentioned the secret tunnels he used to travel to Iraq or "meet important people." But the images, for all their violence, only showed an explosion. No faces appeared. There was no concrete evidence of the executioners or the victims. It was impossible to verify. There were two possible explanations: either Bilel really died, or this was a hoax. Meanwhile, this was the first I'd heard of Bilel and Abu Shaheed's relationship, and I was surprised to hear Shaheed—and others—insist on Bilel's close relationship to al-Baghdadi.

As I've said, when an ISIS fighter dies, a picture is taken of him "in his best light" and disseminated. This photo serves as proof that the new martyr is "now at peace." A lot of people were talking about Bilel, but no one had posted any pictures of his corpse, and that was very unusual. Not even of him in his military garb during his glory years. Rather, the only photo circulating was a screenshot from his SUV video. Strange. Bilel may have been using his army to fake his death. Or maybe he'd been executed by his own people for talking. Either way, I was in trouble. If the terrorist organization had discovered that Mélodie was a fabrication, it would suspect a cop or journalist. The article was coming out in two days. It had already been sent to the printer. There was no stopping it. No! I was letting my imagination get carried away. I tried to reassure myself: Bilel had revealed compromising information, but none of it had been a state secret. The horrors he'd divulged with such relish and detail were already partly known. I followed Lou's advice and turned off my computer. Feeling helpless and confused, we went into my room. Living there was a strange return to my past. It was especially overwhelming to experience the fallout of this story in the place where I'd known some of the best and worst moments of my life. It was disorienting. My reality felt suddenly—and radically—changed. I kept wondering to myself: *What the hell is going on?*

My phone rang every five minutes. It was ironic, considering he was supposedly dead, but that was the night Bilel really got under my skin. And it wasn't over. I wasn't answering anyone, and word had gotten out that Lou was with me. They started harassing her. When my editor in chief called back, Lou panicked and passed me the phone. My editor told me that Bilel wasn't actually dead.

Have I mentioned my head was spinning?

. . . .

David Thomson had rescinded his tweet. I needed to contact him. I wanted to know the truth. My editor offered to speak with him while I untangled fact from fiction on the Internet. I spent the whole evening walking back and forth between the kitchen (the only room with a reliable Internet connection) and my old bedroom. Lou and I made a ton of phone calls, scanned the Internet, and checked the AFP wire. David Thomson kindly called after he heard about my situation. He'd only removed the tweet because Bilel's family had asked him to do so. David hadn't known that they followed his publications. He'd agreed to their request out of respect but, he confirmed, the jihadist was dead. There was no doubt. He had a very reliable source in Syria. Okay . . . Now what?

Lou was reticent to leave me alone with my confusion, and she didn't go home until late. I'd talked to a lot of people that night, and after my friend left, I couldn't get their various hypotheticals out of my head. I didn't know what to think, so I took a sleeping pill to put an end to the day.

Tuesday

My telephone woke me early. The calls were endless. Everybody had the same question: "So is he dead or not?" Which was inevitably followed by: "Does it bother you that he had a wife and kids?" That was the least of my concerns. I barely answered. Besides, I'd already known about his family.

Approximately two weeks before, a woman had contacted Mélodie on Facebook. Strangely, she had been able to send Mélodie private messages even though they weren't "friends" on the social network. Typically, when you try to contact a person who isn't on your friend list, the message gets routed to a spam folder, which no one ever checks. The only people with the power to send private messages to those who aren't on their list work either for Facebook or for the state. Representatives of the latter hide behind fake profiles and contact people in the service of an investigation. At the time, the interior minister's plan to combat the recruitment of French citizens by the Islamic State had only recently been implemented. Mélodie had been updating her profile with posts about al-Sham when a woman named Fatima began sending her messages with strange questions.

As when Mélodie wrote to Bilel, the number of spelling mistakes was staggering. Strangely, however, her grammar was impeccable. She called me "my sister" and said she lived in Tunisia. She claimed to be twenty-eight.

Okay, but what did she want from Mélodie? She immediately mentioned Bilel. What a small world. As fate had it, Bilel connected at the same time, in the middle of the afternoon, and sent Mélodie a string of hearts. "Why are you talking to me about Bilel?" Mélodie asked Fatima. She replied that she'd been planning to go to Syria, but she wasn't sure anymore, since she'd discovered that she was only Bilel's second choice. He was going to marry Mélodie first, then Fatima. Polygamy disgusted her. "It's too bad," she said, "because he's really hot! But he knows it and he uses it to his advantage." Annoyed, Mélodie asked Bilel about Fatima. He replied with a series of "ROFLs" and said not to pay attention to "jealous sisters." I asked Fatima how she knew and why she was telling me. She acted embarrassed. She claimed she was sick and wasn't sure if she'd go through with her jihad. Besides, she said, going to Syria was dangerous. Her tone changed and her spelling suddenly improved. Bilel was dangerous, she warned, and Syria even more so. She was abandoning her jihad, and I should follow her example. For the first time in this adventure, someone was finally telling Mélodie the truth about Syria. Mélodie asked what had made Fatima change her mind so quickly. Only twenty minutes before she'd said she wanted to marry Bilel and go to Syria to be with him. Fatima dodged the question and asked if we could speak over Skype. She seemed genuinely worried about me, repeating my name several times and discouraging me from going to "hell on earth." Before replying, I wanted to "torture" Bilel a little bit. Mélodie was going to start their first fight. Discord adds spice to the beginning of love stories. Raising her voice, Mélodie told Bilel she was disappointed. Meanwhile, the woman from Tunisia kept sending horrible messages about him.

"LOL! I bet she's from Roubaix.* She's just messing with you because she wishes she could be you. You're going to be Abu Bilel al-Firanzi's first wife."

Roubaix? Bilel—Rachid—had always claimed to be from Paris. I turned my attention back to Fatima. Mélodie promised they would Skype, but she wanted to know one thing first: where was Bilel born? Fatima didn't know, but she could tell me that he was wanted by the authorities and already had three wives. Two were converts, like Mélodie, including a girl of twenty. His first wife was thirty-nine, Muslim, and French. Whoa! This was an interesting day! I pestered Bilel a bit more. Without realizing it, he revealed details about his other identity: Rachid. Those details would turn out to be very useful, and not just to me. . . . Bilel continued to deny having other wives, and claimed he'd never married.

I was at the magazine, but I really wanted to go home, where I could have a real conversation with Fatima over Skype. I hadn't brought my veil with me to the office that day. I had suspicions as to her real identity, but it was just a

---

* *Translator's note: Roubaix is a small city in the north of France.*

feeling. It was better to be safe and wear a veil. By the time I got home and dressed up as Mélodie, forty-five minutes had passed, and Fatima's Facebook account had disappeared. The name at the top of one of our messages had been replaced with this inscription: "unknown user." I never heard from her again. Was she a good, if confused, Samaritan? Or was she an avatar for the authorities? I never found out.

I received calls all morning; notably from the magazine. I took my dog for a walk and screened most of them. The editor in chief called at noon. She thought I should leave Paris before the publication of the article. A few hours later, I was traveling with her, gazing at the reflection of clouds in my window. She was taking me to stay with some "close" friends. The location was breathtaking: olive trees, pines, and endless green instantly lifted my spirits. I could breathe here. Everything—inside, outside—was so big, so majestic. The property seemed unreal. My hosts didn't know me, but they welcomed me with kindness. I went for a walk outside, passing chickens, donkeys, and horses. I was accompanied by dogs and a toddling, curly blond *bodyguard*. Nothing seemed serious here. And that felt good. The little girl woke me up every morning at dawn, just a few hours after I managed to fall asleep. When I opened my eyes, disoriented, I found her impish face staring into mine. She was a mischievous child, innocent and pure. A stark contrast to the month I'd just lived. Still, even in this paradise, I felt as if I were looking down on my life. As if my body and mind had become detached.

I owed my hosts an explanation of why I was suddenly staying with them. So I told them. Then I left to feed the horses, before traveling through the labyrinthine Internet looking for clues. I went to see the chickens. Then I answered more questions. And so on . . . My mind swirled with thoughts and emotions: death, Bilel, the future, the present, moving, the little girl, the animals, the guardian angels I'd just met, my family and friends who were far away. Nothing made sense. It was all a mess. This merry-go-round was turning at full tilt. I could jump from my wooden horse and be free of this spinning ride. But I stopped myself. Or something was stopping me.

Eight months later

I wish I could offer a moral to this story. But how can a story have a moral if it isn't finished? More than ever, I feel the threat of danger. It's invisible, unforeseeable, and constant. Or perhaps I'm imagining things. I've asked myself so many questions, which have gone unanswered. If I had to sum up this experience, I would borrow the words of Franklin Roosevelt: "If asked to classify human suffering, I would do it in this order: sickness, death, and *doubt*." The German philosopher Nietzsche posits that certainty, not doubt, kills. In my case, it's the opposite. My uncertainties and the consequences of my actions have thrown me into a mental prison, and only a reality clad in intangible convictions could free me. For the past eight months, since May 5, 2014, my life has been an eternal recurrence. I stopped counting the number of statements I've given to various branches of the police when it reached 254. I've never sought them out; the authorities have always come to me. The DCPJ and an antiterrorist judge also asked to hear my testimony after my real identity started appearing in a number of their files.

Mélodie wanted to help girls like her, and today Anna is paying the price. The authorities, fearing the terrorists could trace my address and my identity, have twice asked me to change my phone number. I don't live in my apartment anymore. "For my safety," the periodicals where I typically publish articles on this issue have banned me from working on the Islamic State and its networks. Drastic safety measures have been implemented at my workplaces. Meanwhile, young girls in burqas, accompanied by much older men, have started asking strange questions at the reception desk. The threats have also gotten more serious. I had to shut down Mélodie's Facebook account, but I can still access her Skype account. The authorities asked me to keep it open for various ongoing investigations. We could also use it to keep an eye on threats toward me. I don't check it very often. Sometimes when I do, I'm greeted by terrifying messages. They started this summer, when I was on assignment in South America. Someone claiming to be Bilel's wife started sending intimidating monologues filled with insults. Her recurrent, nagging question: "How does it feel, bitch, to write a story about a terrorist and fall in love with him?" It was a strange interpretation of events. Was this really one of Bilel's wives? Or was Bilel himself still alive and torturing me? You've probably guessed my answer: I don't know.

Multiple police branches have classified Rachid X., known as Abu Bilel al-Firanzi, as "alive." They have no proof of his death, but they do have a thick file on him. At first they weren't able to identify him. Then I mentioned Roubaix. That allowed them to trace his identity. They'd lost track of him after his departure for Syria. Before that, he'd committed a number of small crimes, from theft to armed robbery. He'd been judged and charged several times in absentia. In 2003 he became an active member fighting against the American invasion of Iraq. That's when he met Abu Bakr al-Baghdadi. Then, between 2009 and 2013, after long trips to Afghanistan (where he perfected his guerrilla techniques), Pakistan, and Libya (at the moment of Gaddafi's fall), he returned home to Roubaix. Without anybody's knowledge. He reappeared on the radar in late 2013, when he was spotted in Turkey. He does in fact have three wives, aged twenty, twenty-eight, and thirty-nine. They're all with him in Syria. He is the father of at least three boys under the age of thirteen. The two eldest are already fighting on the front in Syria. Bilel is closely linked to al-Baghdadi. And to Souad Merah, the Toulouse shooter's fanatic sister.

I've never had direct contact with Bilel again. Recently, when I was on assignment on the other side of the world, a journalist friend called to tell me he'd learned there was a fatwa* against me. He said his source was "a hundred percent reliable." This was the nth time I'd heard that. I'd felt afraid sometimes before, but I'd never felt like I was being followed or watched. Still, I knew this friend wouldn't have said anything if he wasn't sure. So I spent hours searching the Web, forgetting my current job. After a while, I found a video about me. It showed me wearing Mélodie's veil on my couch. It was taken, I imagine, by Bilel. There's

---

** A religious opinion given by a mufti, a jurist who interprets Islamic religious law. The term has become commonplace, and is used to signify a call for revenge by Muslim Sunnis against one or several persons.*

no audio, but it does include cartoon characters of a devil and bilingual, French and Arabic, subtitles. I've only seen the video once. I don't think I'll ever watch it again. But I remember every word:

> My brothers from around the world, I issue a fatwa against this impure person who has scorned the Almighty. If you see her anywhere on earth, follow Islamic law and kill her. Make sure she suffers a long and painful death. Whoever mocks Islam will pay for it in blood. She's more impure than a dog. Rape, stone, and finish her. Insha'Allah.

You decide: which is worse, doubt or certainty?

On June 6, 2014, ISIS, led by Abu Bakr al-Baghdadi, officially launched its first offensive in Iraq, striking the second-largest city in the country, Mosul. After four days of intense fighting, the terrorist organization gained possession and imposed sharia law.

On June 29, Abu Bakr al-Baghdadi proclaimed himself caliph of the Islamic State, changing his name again to become Ibrahim, one of the five most important prophets of Islam. Although he claims to be the leader of all the world's Sunni Muslims, only other terrorist groups have pledged allegiance to him. Most of the world's Muslim authorities do not recognize him as such.

On August 8, President Barack Obama authorized the first air strikes in Iraq.

Since September 24, an international coalition of at least twenty-two countries, led by the United States, has been waging an air offensive in Iraq, and more recently in Syria, against the Islamic State.

Today, it is estimated that 15,000* foreign fighters have joined ISIS's ranks since 2010. They come from eighty different countries, including, according to official numbers, 1,089 French nationals, from 87 national districts. One hundred twenty have returned to France. Forty have died in Syria.

Unofficially, fighters in the Islamic State's organization number between 35,000 and 45,000. The number of French people in the organization is thought to be at least twice as many as stated in the official numbers.

I never heard from Lola again.

Vanessa gave birth to her baby in France, where she currently resides.

Chaquir Maaroufi, aka Abu Shaheed, died "in combat" on June 1 during an offensive in Deir ez-Zor, in Syria.

Abu Abdalah Guitone, aka Guitone, was killed on July 25 on division 17's base in northern Raqqa, ISIS headquarters.

Hamza and his family never tried to contact Mélodie or her family again.

Abu Mustapha, a close friend of Abu Bilel, blocked Mélodie from accessing his accounts.

Rachid X., aka Abu Bilel al-Firanzi, is still considered to be alive by France's various departments of domestic and foreign security.

---

* *Most recent source: the* Guardian, *November 2014.*

About the Author

Anna Erelle is the pseudonym of a thirty-two-year-old French journalist. She was investigating the disappearance of other young people, men and women gone to Syria, when she first went online under the avatar of Mélodie in order to learn more. She has written for major newspapers and magazines in France. *In the Skin of a Jihadist* is her first book.